Solving the Mind-Body Problem by the CODAM Neural Model of Consciousness?

Springer Series in Cognitive and Neural Systems

Volume 9

Series Editor
Vassilis Cutsuridis
Boston University, Boston, MA, USA

For further volumes:
http://www.springer.com/series/8572

John G. Taylor

Solving the Mind-Body Problem by the CODAM Neural Model of Consciousness?

Springer

John G. Taylor (Deceased)
Department of Mathematics
University of London
London, UK

ISBN 978-94-007-7644-9 ISBN 978-94-007-7645-6 (eBook)
DOI 10.1007/978-94-007-7645-6
Springer Dordrecht Heidelberg New York London

Library of Congress Control Number: 2013956004

Printed on acid-free paper

Springer is part of Springer Science+Business Media (www.springer.com)

I have worked on the material in this book it seems for my whole life. Yet the detailed material – conjectures, theories, experimental support – presented in the book has only been gathered together and developed on a solid scientific basis in the last 15 years at most. The book is meant to be a message for all mankind, to allow them to begin to know who they really are. But it is especially for my grandsons Jack and Joe, so that when they are ready they can find out what 'Grandpa London' was trying to do as he watched their wonderful growing up.

I dedicate this book to them so that they do not have to go searching in the highways and byways to find out what life truly is in scientific terms. I hope they can use the contents in the book to have lives full of the understanding, given here as a launch pad, although I am sure they will find their own way forward in understanding life. If anything, the results presented in this book may act as part of the background for their search.

Foreword

Consciousness, also known with other names such as 'inner self', 'soul', and 'I', is one of the great mysteries throughout history and certainly of our time. Many people of different backgrounds, coming from religion, philosophy of mind, literature, psychology, psychiatry and other fields of humanities and science, tried to provide answers now and in the past. Recently, due to the amazing technological and scientific advances of the last two decades, neuroscientists have shown an increased interest in the problem. But why is this question so important? The reason is that the question lies exactly on the interface of our material bodies (Body) and our immaterial spirits (Mind). What is exactly the relation between the two?

The above question accepts a range of answers covering conjectures such as that the Mind is simply the emergent effect of the normal operation of our Bodies or that the Mind depends on the operation of the Body but at the same time it is a somewhat independent entity from it or that the Mind is an entity that exists independently from the Body and so happens that sometimes Mind and Body co-exist in the same person during his lifetime. Indeed the nature of consciousness is a complex question which becomes increasingly important to answer. The motivation is not only epistemological but also scientific and technological. From a scientific perspective we want answers to questions such as: what is the evolutionary advantage of having consciousness? Is it a uniquely human characteristic or animals have it too? Does consciousness require an embodiment or could it exist without a body and the corresponding sets of external and internal stimuli? From a technological perspective, one wants to answer questions such as if intelligent robots and similar machines need to have consciousness for performing better in their environment; if self-awareness is a necessary condition for achieving intelligent systems comparable to human performance or if there is a suitable test for testing consciousness of a machine like the well known Turing test for testing intelligence.

One has to ask oneself if it is reasonable to expect complete answers to the above questions. So far, multiple methods of investigation have been used including introspection, religious experience, meditation to 'pure consciousness states', and recently a great effort has been put forward by the scientific community to explore the issue. During the last 25 years, many neuroscientists, neural network modellers, psychologists and related branches of brain and cognitive science have been trying to apply the scientific method to the problem. Will this work? We do not really know. The situation is reminiscent of the exploration of the nature of things such as the light, the fire and the gravity. It took many centuries to come to verifiable answers and abandon old perspectives for the aforementioned physical phenomena. The question of consciousness goes deeper. If a system, like the human brain or the Internet, becomes complex enough, under what conditions can it show signs of consciousness? Can the scenario of the *Terminator* science fiction movie come to reality? Is there a process shaping the evolution of consciousness like Darwinian evolution shapes our bodies?

So, how can we come to any sort of answers using the scientific approach? The typical process for studying consciousness is that of data collection, from patients and healthy individuals, of building models, of simulating models and making predictions and of searching for overt (directly observable in behaviour) and covert (internal brain monitoring and observation) proof that supports the predictions of the models. While each step has great difficulties, the biggest challenge is the model construction step. The building of a model is a complex, laborious and difficult task often taking place over many years while the investigator collects, analyses, reflects on relevant data and adapts his thinking in light of new evidence.

I remember vividly a conversation with John Taylor, on our way back to London from a conference in Turkey on 2002, where John was explaining to me his ideas about consciousness and how he planned to develop a research programme to investigate the problem further. This was a time where his ideas took shape in the form of the CODAM model. His fascination with the problem was much older, starting from the 1970s. I recall his drive and enthusiasm and his firm belief that progress can be made through the use of the scientific method. Many aspects of the problem were discussed in that flight including the question if one had enough computational power available to simulate a global model of the brain, with enough detail, in order to observe its states under different sets of stimuli coming from the outside world.

Many years have passed since that flight and John's ideas have further evolved while in parallel he was developing models for the components of the global brain model. He emphasised the importance of attention as the top-level control system of the brain. He considered attention as the gateway through which we can probe the consciousness question and he even believed that systems that do not possess the faculty of attention could not display consciousness. This is still, of course, an open issue and more data are needed to answer the question in a conclusive manner. Still his conjecture, on the importance of attention for consciousness, is a step forward in identifying the necessary and sufficient components that should be in place for consciousness to appear.

We have worked jointly in a research project, during 2004–2007, where we tried to construct a reasoning robot which could solve tasks that required the use of tools for their solution. These tasks were similar with the tasks put forward in animal experiments with chimpanzees and New Caledonian crows. A direct comparison of the behaviour between machine and animal led us to insights as to the nature of the processes that potentially support consciousness. This project provided the means for John to advance further his thinking and to work out in detail models of older and newer components of the global brain model such as the attention system, the reasoning system, visual processing and others.

In this book you will find a panoramic view of the problem and he will masterfully guide you through the quest for answers. Consciousness is not an easy issue to explain or attack. Introspection reveals a sense of 'nothingness' when we search for our inner self. It is yet unknown if more advanced technology (e.g. brain scanners, better computers, etc.) will provide conclusive evidence on the question. This can cast doubt in the mind of many people as to the usefulness of the scientific approach. Still the 'battle' has only begun and it counts only two decades of systematic investigation. Computational models of consciousness have started appearing in the scientific literature.

Having the opportunity to know John as a teacher, mentor, colleague and friend, for a considerable amount of years, I believe that the first pillar of his legacy is his drive to always focus on the right question, the important, the central one and not be sidetracked by the peripheral issues, the details and the process. The second pillar is his holistic view as the approach to be used for the solution of the important questions. Initially a theoretical physicist, he had to master subjects, such as mathematics, neurobiology, brain science, philosophy, cognitive science, artificial intelligence and others to become the integrating generalist who had the tools and knowledge to attack such an important question. As an accomplished scientist, a foremost expert in the fields of Neural Networks, Computational Neuroscience and Theoretical Physics, a true Renaissance man, he is the right guide for a tour of the Mind-Body problem. He provides an engaging view to the mysteries of the Mind and we are lucky to share his insights developed over a period of 45 years of science and thinking about the nature of consciousness. I am confident that you will enjoy this trip to the wonders of the Mental Universe.

John is no longer with us since 2012. His legacy will guide the next generations of enquiring minds to their search for answers in the great mystery of the 'I'. He was an inspiring teacher and a wonderful friend. He will always be remembered fondly from all the people who knew him.

Lead Scientist Dr. S. Kasderidis
NOVOCAPTIS Cognitive Systems & Robotics
Thessaloniki, 2013

Preface

The main purpose of the book is to present, through material based on a collection of relevant past scientific papers by the author, a complete development of the CODAM neural network model of consciousness, first introduced by the author in 1999/2000. This model allows both for a scientifically based (and scientifically supported) understanding of consciousness as well as providing a solution to the outstanding Mind-Body problem, a problem which has been argued over (sometimes lethally) for several thousand years. As such, this is a very ambitious book, but one that can carry the weight of the scientific community: the book is based on material taken from already rigorously refereed scientific papers in scientific journals, all written by the author (some with colleagues), on the subject of consciousness and the Mind-Body problem, and forming a coherent whole. To make the book more complete there are also various chapters that extend the understanding of consciousness provided by CODAM itself into a more detailed discussion of the overall Mind-Body problem.

In particular, there is an extra chapter completing the lengthy journey undertaken by the great Charles Darwin to understand the variety of life through evolution: this particular chapter provides a detailed explanation of the evolution of human consciousness (an aspect Darwin was either too frightened or too exhausted (or both) to contemplate in any detail), along with scientific support for the approach (based as it is on the author's CODAM model of consciousness). It is that chapter on consciousness evolution which completes the journey started by Darwin (and his co-evolutionist Wallace), and was continued by Crick and Watson with their proposed double helix structure of DNA. That journey is now completed by following the evolution needed to create CODAM-like models of consciousness.

As well, the understanding presented in a further chapter allows for a more detailed discussion of the whole area of animal consciousness, itself very contentious but highly relevant to animal welfare. This has become of increasing importance fed by the enlargement of our knowledge on animal intelligence, even leading to a suggested Charter for Intelligent Animals, which is taken to include Ceteceans (dolphins, whales, porpoises) and should also include the Great Apes.

At the same time, the main body of the book contains chapters on the mental disease of schizophrenia as seen through a CODAM-based magnifying glass, and on meditative states. The many cases of schizophrenia are very important to be understood, and we follow the psychiatrists Sass and Parnas in showing how damage (from a variety of sources) can cause there to be distortions to experience coming under the general description of 'loss of 'I'', a very important complaint in schizophrenia, especially for youngsters in the early prodromal stage of the disease.

The meditative states (the so-called 'pure consciousness experience' and others) are also able to be shown to be understandable in terms of meditatively-based modifications of the CODAM apparatus in the subject's brain. The implications of this for various stages of meditation and the related meditative states (pure consciousness, oceanic consciousness and beyond) will be considered as part of this CODAM-based investigation.

There is currently no other publication (book or journal paper) which gives a general but scientifically-based approach to consciousness powerful enough to lead to what has been called by Western phenomenologists the 'inner self' and its ramifications, highly relevant to explain the vast range of human experiences, as is described in the material in this book. The gap thereby being bridged can be seen to arise mainly from the divide that exists between philosophers of mind and the neuroscience community. The first group considers mainly features of the mental world, the neuroscientists on the other hand look in detail at the structures and dynamics of the brain. This book is therefore to be regarded as a unique bridge between these communities. This bridge is very important for allowing the enormous weight of theorizing on the nature of mind to be brought to earth and put under the probing gaze of science.

It allows in particular for understanding ourselves scientifically as individuals as well as in interactions between each other in society. However this book tries to present mainly the scientific basis of CODAM and related scientific features; implications of the CODAM approach for society more generally are covered in my 2006 John Wiley book *The Mind: A User's Manual*, giving a truly human direction to the whole enterprise. Thus this latter book can be considered as a companion to the present one, exploring the further implications of the CODAM theory in more human terms. However we first have to consider the scientific theory of CODAM as thoroughly described and explored, especially supported by known experimental data.

The unique features of the book are thus:

- CODAM is the first model of consciousness to be supported by detailed scientific experimental data from the human brain.
- CODAM provides an approach to the evolution of human consciousness over the aeons, thereby also providing a unique understanding of possible animal consciousnesses as evolved on the way to the human variety.
- CODAM provides a framework in which to understand the whole area of human meditation experiences, including pure consciousness and more complex meditative states.

- CODAM provides an understanding of a range of mental diseases, especially the key one of schizophrenia.
- The ultimate application of CODAM is to understand humans in society from the unique viewpoint of the brain basis for human consciousness. This is given in terms of their emotional and intellectual needs by a developing CODAM-based brain (as discussed more fully in my 2006 Wiley book *The Mind: A User's Manual*).

It should be pointed out that this book is not an easy read. Parts of it may be, but the crucial chapters of 5–15 will require careful reading (possibly more than once) to get the full picture. Furthermore, the main thrust of the book is based on an engineering control view of the brain, and most especially of its faculty of attention. Yet the content is heavily dependent more generally on neuroscience, the theory of neural networks, computational neuroscience, basic ideas from psychology, and even some components from mathematics and physics. But then we must expect that such a broad brush is necessary to properly tackle one of arguably the most difficult problems of the Universe: the Mind and its explanation in physical terms.

Chapter 1
The Story of Consciousness

1.1 The Beginning of the Story

We look out into the Universe to attempt to understand its nature. In the process we have discovered strange and disturbing objects like black holes, surrounded by an event horizon to shield us from their impossibly bizarre centre, where matter magically disappears. Our search in the heavens also involves a search for ourselves, and how we fit into this bizarre material universe we are discovering. So at the same time we are searching the heavens for ourselves and our own souls (whatever they are), thereby hoping to make sense of our inner black holes. For we have something equivalent to those exotic black hole entities at the centres of our being – they are our central core of experience. We try to find a place to stay equably outside the event horizon (the region around the centre of a black hole from which there is no escape if once entered) of our inner central black-hole-like attractor. For some people it is hard, with them falling ever faster to their inevitable end. But for all of us, however hard we struggle we must eventually turn to look at our own minds, and especially that central inner black hole that seems empty and which we try forever to escape. It is our core 'I', sometimes called our soul. What is this 'I' to which we must face up in order to make our lives complete?

In the process of such a search we are trying to create the complete story of 'I'. This is 'the greatest story never told', and only able to be put together in a tentative version at this stage. But the time has come when the new knowledge from brain science and archeology and from the new powers of computers will allow us to piece together what roughly may be happening in the brains of every one of the seven billion people on earth presently (although that is still expanding rapidly), and the story of how that consciousness began and evolved. Such a complete story should help to explain especially the inner black holes and the apparent vacancy of any 'I' at the centre of experience of each of us.

The emotions, memories, perceptions, decisions and actions of humans are now being increasingly understood on a scientific basis. Mathematical models of how these various systems work have been constructed and proved successful to explain

J.G. Taylor, *Solving the Mind-Body Problem by the CODAM Neural Model of Consciousness?*, Springer Series in Cognitive and Neural Systems 9,
DOI 10.1007/978-94-007-7645-6_1,

many of the phenomena observed by brain scientists. Robot brains have been created that can endow their possessors with powers of reasoning and thinking. On this firm basis further models have been developed of the 'inner self' of humans. In this manner the beginnings of a view as to how the seven billion humans on earth fully experience the world around them can begin to be constructed. The inner self of each of us has now begun to be glimpsed, at least through a glass darkly. At least we can begin to see the event horizon round that inner black hole of the soul. However, as in all scientific endeavours, the story to be told here is still to be questioned by much more detailed scientific analysis over the following years, decades and centuries from now. As in all such attempts science itself has to be questioned as a way to understand the Universe and our place in it. Perhaps it will work, perhaps it won't. But only following the scientific yellow brick road will we discover if science works or it doesn't in the arena of the 'I'.

The dark glass we are looking through at our inner experience lets in enough light to enable us to see the rough shape of the entity that has been called a 'soul' by some, an 'I' by others, an inner self by others. It is these identical components that provide that inner central core of each human's experience of the world around them. It is their event horizon round the frightening inner black hole of the soul. Without a soul a person has no experience – there is nothing it is like to be them. With it, the inner life takes off and makes sense for most. But it is a frightening object to many, as is a Black Hole in space. However by better understanding of the soul's nature we can begin to understand for the first time how inner experience is, for example, distorted in a mental disease such as schizophrenia, where a terrible alteration of the inner life causes it to end up as apparently controlled by others. The disease can also lead to a loss of the sense of 'I', so that the sufferer begins to lose contact with the outer world – it becomes an ever more frightening place. How that works out can now begin to be followed in detail.

There is more to the Complete Story of 'I' than explaining distortions of experience in disease and the normalcy of ongoing experience. The Story Proper involves how it was that man himself came clawing through the competitive jungle of the beasts to so sculpt his or her brain that it evolved to possess an inner self. As the great Charles Darwin (1879) would expect it, the 'I' gave a decided advantage over those animals without it. When the 'I' emerged in the evolution of the animal kingdom will also be considered here. This thereby throws some light on possible animal consciousness and the corresponding rights of animals.

It should also be possible through this understanding of our experience to arrive at a true reconciliation between science and religion. Our analysis of 'I' and its extension to the meditative state termed 'Pure Consciousness' will allow us to explore the various aspects of the latter. In the process we recognize a commonality between religious experiences of the great mystics of the past. Having an explanation of Pure Consciousness thereby allows us to begin to understand the religious experiences of those who are at the bases of many religions – Jesus, Mohammed, the Old Testament prophets of the Hebrew faith, Buddha, and so on. We therefore arrive at the deep conclusion: all faiths are based on Altered States of

Consciousness. Given these states provide solace to many people, religions could thereby be drawn together by this commonality, and be so developed as to provide even deeper solace by the enhanced glimpse they provide each person of their own consciousness, and especially their own 'I'.

To round out the Complete Story we consider if it is possible that future genetic engineering might modify our inner self so as to make it ever more efficient and increasingly superior. Are there genes that could be tweaked (or added or removed) to produce a race of super-intelligent humans with super-consciousness? Also what could alien consciousness look like in our terms? Some of such possibilities will be discussed to consider the possible end of the Complete Story of 'I'.

This then is a unique book – the Complete Story of 'I' – never considered before and only now able to be told, although possibly only in a preliminary version. It is a thrilling story and holds for everyone a fascination that should let them see their own life in a new light.

1.2 Fitting into the Universe

In the very beginning was the Big Bang. It was in fact very, very big, since it generated the whole Universe we know of, with its hundreds of billions of galaxies, each with their hundreds of billions of stars, and many of those with their planets swirling around them. But since it used up all the space (in fact it generated its own space) there was nothing outside it. So the Universe began from a point, and has been expanding ever since over billions upon billions of years (13.5 billion or so, to be more precise). Certainly the Universe did not start in 4004 BC, as was confidently proclaimed by a certain Bishop James Usher in the seventeenth century, after his careful computation from the Old Testament of the expected time since man was on earth by counting all the generations listed there. In fact it is now known, from estimates of the proportions of various radioactive elements present on the earth's surface, that the earth is at least several billion years old. So the Universe is certainly much older than expected from the Scriptures.

It was not until 1929 that the American astronomer Edwin Hubble directly observed the galaxies in movement away from the earth by using the biggest telescope then on earth. Since that time we have had to acknowledge that we live in an expanding Universe. To explore this further, Hubble drew the trajectories of the galaxies back as far as he could and discovered that they met together in a point 13.5 billion years in the past. This really drew a line under the Big Bang – scientific proof of its happening. And the Universe itself was 13.5 billion years old. And it is still going strong. Many more Happy Birthdays, Universe!

It was not for some time after the Big Bang that our solar system came into being, after the first generation of galaxies formed, only half a billion or so years after the Big Bang. These then began to dissipate, in the process spewing out their heavier elements (even up to uranium) which were sucked up as our earth condensed from the debris round our fledgling sun somewhat later. We could then use uranium and its heavy colleagues to heat ourselves, and kill each other ever more efficiently by atomic bombs.

These processes of solar birth and death, both early and late in the Universe, are still going on all the time. We can look up in the heavens and see stars in the process of forming by collapse of the dust clouds that will ultimately be their constituents. Some stars have quite short lifetimes, since they are very heavy and use up their energy to blow themselves apart relatively quickly. Other stars, like our own, live for about 10 billion years. Ours has another 5 billion or so to go before it too blows itself apart as a supernova.

More recently we have discovered the Black Hole, of which there are many examples around in the Universe. Some are star-sized, coming from single collapsing stars after they have used up the process of fusion generation of power from their initial fuel of hydrogen. But there are also much larger Black Holes, thought to be at the centres of most if not all galaxies. Our own Milky Way has one of those beasts at its centre. These Black Holes were very likely created at the very early stages of the Big Bang, and have been trying to eat up the surrounding stars in their galaxy ever since. In the process they have been expanding, to become ever more obese. No diabetes for them, just getting bigger and bigger and more superior! Round each of these Black Holes are event horizons, which allow their centre points to be hidden from us: for at the centre of each Black Hole, matter is magically disappearing. Where it turns up, if at all, is unknown although alternative Universes have been suggested. The Black Hole is indeed a bizarre and worrying object out there.

But where can we ourselves fit into these enormous processes? These astronomical machinations are superhuman, or more correctly put, they are inhuman. The Black Holes are the most worrying, since they seem to be places where everything is totally annihilated (Taylor 1973). However why should there be any human-orientation out there in the Universe? It would be as if we – all seven billion or so of us humans here on earth – were not needed. We are irrelevant. We could even just have been a dead end or, forbid the thought, a mistake. How are we, with our apparently immaterial minds, able to fit into this maelstrom of matter being spewed forth from orifices and from around the event horizons of those most dangerous of all, the collapsing stars going to make Black Holes? Do our minds set us apart from all this material stuff, so we are from a different Universe and happen to have latched onto this one as a happy medium for our actions stemming from our thoughts and consciousness? In other words are we having a free ride on this so-active material Universe, full of dangerous Black Holes ready to swallow us up, but not being able to do so because we are immaterial in essence?

Instead, we could just be part of the same old matter spewing away around us, having evolved as part of Darwin's great theory of evolution, and going from 'dust to dust' as the Scriptures would have it. Then we would be vulnerable to the Black Holes and exploding stars, as well as to the desecrations of our own earth-space that we are presently making because there are too many of us and we are all too greedy.

Our minds seem to obviate that danger. They do indeed seem to be able to stretch out to infinite distances and to allow us powers of imagination that take us into new spaces and faraway places. They allow us to think we can live forever. Our minds must thereby be immaterial, whatever that means, mustn't they? But our minds also seem to contain an amazing structure which I have already called our own inner Black Hole – as our central inner self or soul. It is there, but if we try to probe it the mind apparently disappears. There is nothing there at all. This was underlined by the French philosopher Jean-Paul Sartre in his magnum opus "Being and Nothingness" (Sartre 1943). The 'Nothingness' for Sartre turns out to be exactly our inner self, our 'I'. So we are naturally frightened when we discover the mental inner Black Hole of nothingness at the centre of our Mind.

Some of us turn away from it and distract our mind so as never to dwell on it – using the ever more prevalent electronic devices such as iPods, iPads, i-peds, or whatever. Others try to dive into their own Black Hole by meditation – to catch the devil by its tail, so to speak, and defuse it or put out its fire. Many try to defuse it by the ritual of religion, passing over to the priest or guider the responsibility of thinking about such a dangerous thing. Others dive into the world of jobs, of work or of politics. The 'soul of the party', for example, expresses exactly the way that people will project their inner self into an exterior activity and thereby defuse what they see as a nothingness that threatens their continued existence,

But there are numbers of humans who are more down to earth and who say 'rubbish – we are all made of that same matter that everything else is composed of up in the stars and distant galaxies. Of course there is 'dark matter' and 'dark energy' amongst that matter which we do not understand as well as we would like. But undoubtedly, they say, we will gain that understanding in the fullness of time'. So we should, according to these hard-nosed scientists, get on with understanding our minds as we do our bodies and the rest of matter. That is a distinctly alternative and more scientific route to try to understand our place in the Universe.

1.3 We Are the Hollow Men

But we are the hollow men, like an Easter Egg with no middle cream, or like a hollowed-out wooden doll, as I have already suggested. Think of a sequence of Russian dolls, each doll inside another doll inside . . . – each doll being hollow. We can even look inside ourselves and try to find where our 'inner self' is – where is the 'I' we constantly use when we talk and think. 'Yes, now I think that . . . ' we hear on all sides. Politicians constantly use their 'I' word. But to what are they referring? Is it to a little man or woman sitting inside their head, directing their thoughts? But then who directs the thoughts of the little man or woman? Is it the case that there is another even smaller man or woman inside that one? But that way leads to madness – an infinite regress, with little men inside smaller men inside even smaller men . . . without end So who is there to be an 'I'? Is there nobody? But then we are hollow indeed.

For many years I (sorry about that but I – again – cannot avoid it or rather 'I' cannot) thought I was a hollow man. TS Eliot, the great American poet, was apparently correct, when he wrote in his magnificent poem 'The Waste Land' (Eliot 1922) that

> We are the hollow men. We are the stuffed men. Leaning together. Headpieces filled with straw.

Thus he wrote and the next day the whole world woke up to being full of hollow men. His phrases resonated – very likely in the space inside the head in which the hollowness resides. If you look into the brain there are, it is true, holes – the so-called 'ventricles'. But they are just that – holes – certainly not black holes. The ventricles are definitely no place to fit an infinite sequence of ever smaller hollow men. Nor are there any other places in the brain where it is obvious that our hollowness could reside. So our 'nothingness', as Sartre called it, does not seem to have any physical site in our brains in which it could reside. Perhaps it is our nothingness that is immaterial after all? Or even does not exist, as some thinkers on the subject would have us believe.

I have tried to find out for many years where this hollowness resides or how it works or even exists. You might say that hollowness is exactly, as just mentioned, what the French philosopher Jean-Paul Sartre wrote about in 'Being and Nothingness' – it is the 'nothingness' that to Sartre was the essence of ourselves (Sartre 1943). It has no content, no way of putting a finger on it, except to say that it is the opposite of 'being' – it is 'not-being'. He expanded on that in his great book and laid out what appears to be a complete model of the mind.

But how can something that is nothing be something as important as 'I'? The 'I' is the essence of me – both to the man in the street and to the politician. Without an 'I' to be the source of tax-cuts for the man-in-the-street, or for the striker the source of their wish to bring down the management – then the world would cease to go round. Commerce would stop. Bank bonuses would cease. So maybe after all it is a good thing to have no 'I' – all would be steady and constant.

The Buddhist religion suggests that an 'I' gets in the way of our proper living. Some members of that faith meditate over many strenuous years to get rid of their 'I', whatever it is. Being 'I'-less in society, they claim, allows one more easily to suffer the 'slings and arrows of outrageous fortune', in the words of the Bard. It essentially allows one to withdraw from the real give and take of argument and from the pain of living. So it is not necessarily a good thing. To live half a life, with lessened pain and ecstasy is not necessarily living a real life at all. It can only ever be at most a half-life, if that.

1.4 Where Am 'I'?

So we get back to 'I'. But what have we got back to? That is the question I have wrestled with all my life, from being a small boy trotting besides my father as we 'swam' through the Yorkshire Dales (or so it seemed in our walks in the constant rain).

He would spout about the great thinkers of the past who had tried to bring 'I' to understanding. The names of Plato, Socrates, Aristotle, Descartes, Bishop Berkeley and many more dripped from his lips as the Yorkshire rain dripped down my neck. My father believed all was mind. 'It's all in here' he would proclaim beating his breast. 'It's all will, intellect and emotion' he would say, again and again, until I was almost completely brainwashed into seeing the material world around us – including the heavy Yorkshire rain dripping down my neck – as purely a construct of our minds. It all sounded so reasonable, since my mind was at the centre of my world. Why not spin the material world presently pouring down my neck out of that mind?

But that material world has proved far too complicated to construct it out of just will, intellect and emotion, whatever they were (as my father had manfully tried to do). I soon began to realize the complexity of those three innocent-sounding entities. At the same time I found out that the material world was itself full of many surprises. So I grew up to be a theoretical physicist and worked on elementary particles and superstrings, where the surprises came in thick and fast to my colleagues and me as we worked. Very, very different from those three simple little factors of mind my father had bequeathed to me, of will, intellect and emotion.

But then I came back to the brain and mind with a vengeance with the advent of ever bigger and better computers and of the creation of new machines to probe into the brain without cutting it up. The big computers are well known – and the progress of their improvement in speed and size is also well-known. We all now have on our desks or in our mobile phones computing power unimaginable a few decades ago. That power is getting close to that of our own brains (although we have not yet worked out how to employ it as effectively). The brain imaging machines – magnetic resonance and positron emission tomography machines able to localise the activity of a particular piece of brain tissue to within millimetres – are also increasingly in front of the public eye. Pictures of brains at work appear in increasing numbers in the newspapers and on television since they are producing amazing insights into the way our brains control our behaviour. It is these developments that drew me to looking seriously at the brain since the time now seemed ripe to do so with good scientific backing. It was as if the Hubble telescopes of the brain – the new brain imaging machines and the new computing power – were now available to search out the Big Bang and Expanding Universe of the brain. To see the brain in action, at all ages from birth to old age – that would make the universe of the brain the new scientific laboratory to replace those exploding stars and Black Holes and new elementary particles that made such an impact on us in the 1950s and 1960s and helped us to understand the structure of the whole Universe. Now we could turn to the Mental Universe and its analysis by the Mental Revolution.

1.5 In Praise of Revolutions

But getting the Mental Revolution going has not proved so easy. There are many aspects of this inner Universe – the Mental Universe – that have stood in the way of a Mental Revolution occurring at the same speed as that of the Matter

Revolution. This latter was spawned originally by the Renaissance and considerably later developed by the experimental apparatus created on the back of the deadly machines needed for numerous wars. These war machines helped usher in the period of ever bigger particle accelerator machines to probe ever more deeply into the interiors of atoms. They then became powerful enough to let us see the composite nuclei, and then the elementary particles (protons, neutrons and electrons) that made up those nuclei, and thence to the quarks, gluons and leptons of which everything material is apparently made. Now we are in for the next round with the largest particle accelerator ever – the Large Hadron Collider (LHC) – now on-stream in Geneva, to see what makes up the quarks, gluons and leptons, and what the Universe looked like very, very soon after the Big Bang.

Instead of having such relatively plain sailing the mind revolution has had numerous bodies of opinion entrenched against it of not inconsiderable weight: the church ('we have immaterial souls' they proclaim), the humanities ('we are not machines' they plead), the psychiatrists ('we are chemical/genetic machines' to them now) and the man in the street ('we are the hollow men' seems to be their refrain). All has not been as easy as one would have hoped in the attempt to probe the material basis of the mind down to that of its 'I'.

However the data on how the brain controls behavior and inner experience has moved ahead, since it has proved important in attempting to understand how brain diseases could arise and how behavior deficits could be caused by brain injury and to improve treatments to help lessen their effects. So in spite of the well-established bodies ranked against our gaining increased understanding of our minds and brains, the project of brain science has moved forward. Meetings of the Society for Neuroscience have increased in size and importance year on year and bring in 30,000–40,000 attendees annually.

I well remember attending a meeting in New Orleans of the Society for Neuroscience a few years ago. Of many special sessions was one on Parkinson's disease. It was attended by some thousand or more listening to important developments in the understanding and treatment of the disease. Sitting in front of me was an attractive young woman neuroscientist listening raptly to the presentations. I thought 'how attentive she is' but then I noticed her rather excessive head tremor – a near-certain sign of the disease. It was very understandable for her both to find out about the latest aspects of understanding and treatment for its own sake but also very likely to be working on her own disease itself. 'Life's a bugger' I thought, and wished her all the wellness I could (in my own mind).

1.6 Brain Science Moves Ahead

Given this increasing weight of data from brain science something has had to give. The need to understand how these brain activities being observed under a whole host of paradigms could fit together has been the spur. Clever minds, especially those needing answers for themselves or their friends and loved ones, have tried to piece the brain together from this increasing wealth of data. It is not easy – closer inspection has shown the brain as possibly the most complex physical system, certainly on this earth and possibly through the whole Universe. Black holes and their relatives the neutron stars are trivial in comparison. Indeed there is still far too little understanding of the brain. And behind it all stand, to some degree, opposition from the established bodies I mentioned earlier – the churches, the humanities, psychiatry with its wealth of understanding of the mind but turning away from the centre point of 'I' to increasingly hard science to gain credibility, and the man-in-the-street, with his or her wish to get away from it all and forget that apparent hollowness by diving into his or her latest electronic machines.

Part of the problem of the brain and mind is that they are so completely different. The mind does indeed possess a serious inner Black Hole – its 'I' – whilst the brain is just two fistfuls of porridge. I know that there are beautiful coloured pictures of images of the brain at work, under all sorts of conditions. But the colours are added to help understand what might be going on. One can say that really the brain is colourless. So how could it contain or support such a subtle and colourful item as a mind? Especially one with an inner Black Hole somewhere about. I think we have to accept that 'I' definitely exists. Otherwise we ourselves, each of us, does not exist as a conscious being. The nature of 'I' as an inner Black Hole – circling a nothingness at its centre – needs to be considered in more detail, but is initially appealing. It explains our apparent hollowness.

But many of the hollow men and women of this world do not wish, it would seem, to contemplate their hollowness, as I mentioned earlier. They wish to escape from it, using all the increasingly gizmo-laden society to do so. They use mobiles, personal computers and now increasingly iPods, iPads and so on. Interestingly these i-machines all start with an 'i', not an 'I'. It is as if the 'I' of the person can, by use of one or other of these machines, downgrade their 'I' to such an extent that it no longer causes grief. It is no longer in capitals. Their inner Black Hole is no longer threatening to swallow them up.

As for the others in opposition, the church clearly has a deep investment in an immaterial mind. Without that there could be no God or heaven or whatever. Such a possibility – God does not exist – is strongly dismissed by the believers, in spite of the increasing secularization of society. However the use of the ritual of the church – 'repeating after me' – helps to hand the unloading of their 'I' from father to son or from one believer to another. So there is still much to hold the faith together and to continue it.

Psychiatrists are a strange race. They say they are most concerned about the patient. Indeed the latest approach to the patient through psycho-genetics, on the back of understanding each human's genome, is claimed to be able to improve the ability to prescribe the drugs most appropriate for a given patient, with least side effects for that patient. Such advances are important but are only to be seen as part of the larger and more important need to understand what the patient is experiencing and how that can be ameliorated as well as how their disease can be best combated. The patient should still come first, in spite of the increasingly hard science being brought into the arena by psycho-genetics. But the amazing feature of modern psychiatry is the paucity of interest in the profession about the inner self of the patient. Indeed many recent books on psychiatry have even no, or a very reduced, entry under the index heading 'consciousness'. But that is the 'leader of the gang', so to speak, for the suffering patient!

The 'I am not a machine' cry from the humanities is indeed very important to take note of. But the cry may not be true. It is correct that we each of us have no cogs whirring round inside our heads, as the old machinery would. It would very likely make our heads too heavy to carry on our shoulders. Think of the ancient Greek astronomical machine, the Cythera, full of its cogs and gears! But there are many more sophisticated machines now on the market that are increasingly miniscule. I was involved over several years in an EC Project called GNOSYS (Greek for knowledge). We tried to create a reasoning robot. No cogs whirring round a la Cythera, only software in control of a moving and grasping robot. We succeeded. The GNOSYS robot could reason how to solve the 'two-sticks' problem, for example, in which a monkey (or a crow) in a cage needs to reason out how to pick up a shorter but accessible stick to thereby draw a more distant but more useful longer stick to itself (by the aid of the shorter stick) and thence use the longer stick to draw to itself a grape (or other item of food) otherwise out of its range. Such a task can be solved by monkeys and by New Caledonian crows. Our GNOSYS robot could also solve it. But it had no cogs whirring around in its brain, only electronic processors actively communicating to each other.

There are numerous other machines around that can be said to begin to show a nascent intelligence. It is not that robots are now ready to take over the human race and do all their jobs (a typical worry of many humanists). It is that some of those human attributes, such as reasoning and also creativity (which we can also simulate, as will be considered later) can now be modeled as if they were being carried out by the human brain. Even the neural basis of the emotions is beginning to be understood at a mathematical and neuronal level, as will be described in a later chapter.

Such progress implies that some of these human attributes of the mind – reasoning, thinking, emotions and creativity, for example – can be understood in terms of activities going on in our own brains. But that does not mean that we have understood the 'I'. Indeed I might add to the list of those establishments who stand in the way of progress towards that understanding some of the very neuroscientists who would be expected to be most keen to pursue it. That is because consciousness research has had a bad name for many years, and the very notion of a 'conscious brain' has been of more philosophical interest than of neuro-scientific concern. Consciousness has seemed so private – the ultimate in subjectivity – that its study has been 'outside the pale' of hard-nosed brain science. That branch of science should, according to it practitioners, only be concerned with objective 'facts of the case' and never the inside subjective experience of people. Only recently has consciousness become of real focus as a legitimate part of neuroscience. But that is still somewhat limited in that 'I' never really appears as a topic of research interest. Very few touch it, since the 'I' is either regarded as non-existent (we are the hollow men) or presently too difficult to track down.

So brain science is still in its infancy. Some suggest it will grow up to be strong and healthy, able to explain all of the vagaries of the mind as well as those somewhat simpler (but still very complex) ones of memory, emotion and decision-making. But how can the scientific approach ever begin to bridge that gap from matter to mind? It would still seem that these two are like chalk and cheese, being of a completely different character from each other.

1.7 The Next Move

It is quite possible, in the light of the resistance to finding 'I' which I have outlined above, as well as the difficulty in crossing from matter to mind and vice versa, to go with the crowd and let 'I' find itself, whatever it is. Alternatively 'I' can be allowed to hide itself behind the increasing numbers of electronic gizmos becoming available. That is quite a legitimate position to take, especially if you are a hard-nosed scientist, as I was trained to be. But I have met many similar hard-nosed scientists as I have travelled around the Universities of the world, who say to me, after I have given a talk on consciousness or some related topic: "I went into science to find answers to exactly the questions you have raised" (about our 'I') but they go on to say "but sadly I got sidetracked into X (or Y) subject which it seemed necessary to solve first" That is a legitimate position to be in but one that somehow evades the main question: what indeed is 'I'? We must not put off the day of reckoning when we stand up to be counted in the search for our souls.

But can we avoid getting bogged down in the details of the many sciences related to brain science (and even in brain science itself) by trying to attack the problem of 'I' from the top? In other words can we look at the problem of 'I' from a functional point of view, so as to avoid many of the details of the nuts and bolt of the brain or of the sensors or whatever. These minutiae can distract one, as my scientific colleagues explained to me and as I recounted above. But if we want to take a properly efficient approach we must take account of all that is known about 'I'. We need to expand our view, since we must realize there are indeed bigger fish to fry. These arise since 'I' is actually bigger than just a little man in the head or a personality quirk. What is at stake is nothing less than the human soul. It is that we need to understand, the most important feature of human experience with its great richness. Can it be handled only scientifically or are there features associated with the soul which automatically put it out of reach of science?

But hold on – in any case is science so important here? Or is it just another faith being brought in to help us understand but having its own foibles and dogma? I am often challenged by religious friends who claim that science is just another set of dogma. So what is the difference between religious and scientific views on Life, the Universe and lots of other things? What about other belief-systems, such as communism, fascism or even democracy? Are their associated 'souls' of importance to us, in our search for our own soul? These questions about science and other belief systems must be seriously discussed as part of a consideration of the overall nature of belief and knowledge as we progress in our search to understand the inner Black Hole of the soul. It is a serious quest, one we must treat with great care and thoroughness. And how we justify our progress is clearly of deep import.

1.8 The Empires of the Mind

The search for the soul was beautifully developed in the fourteenth century by the Italian writer Dante Alighieri in his 'The Divine Comedy', which starts with the verses

> Midway the way of life we're bound upon,
> I woke to find myself in the dark wood,
> Where the right road was lost and gone.
> Ay me! How hard to speak of it – that rude
> And rough and stubborn forest! The mere breath
> Of memory stirs old fear in the blood

Dante, as the hero of his book, accepts guidance from the classical poet Virgil so as to avoid the terrible dangers on the route ahead of him. He finally comes to 'see the blessed souls in Paradise' and to converse with them. As Dante stated in a letter written to a patron, the subject of his work was "the state of the soul after death straightforwardly confirmed" So Dante already had accepted the existence of the soul, both in life and after death, and was exploring its possible states – in Hell, in Purgatory or in Paradise. But he lived in the highly religious society of the time, so that was par for the course. We now live in a far more secular society, and need to discover if we really do have a soul at all, and if so what sort of entity it might be.

We are each of us the owner of our inner experience. You do not feel my pain when I stub my foot against a piece of furniture nor I yours. We each live in our own experienced world with, for me, my self – my soul – my inner self – at the centre of my experienced world, and your corresponding soul at the centre of yours. In this way I claim that the soul regains its legitimacy which it has lost in the face of science looking only for observable phenomena. For each of us clearly possesses such inner experience. Each of us has some sort of ownership over our own experience, which is different from everyone else's.

This is an arena which has been thought about carefully by a number of recent world-famous philosophers, one of the foremost among them being the Austrian philosopher Ludwig Wittgenstein. He asked:

> If you tell me you are in pain I cannot ask you if you are sure it is you who are in pain – might it not be someone else in pain? For you are absolutely sure it is you – there is certainty in your mind that you are experiencing exactly your own feelings and those of no-on else.

Of course there can be breakdowns in the brain and mind, such as in schizophrenia, when a sufferer thinks that others are putting alien thoughts into their heads. But let us not consider breakdowns yet – let us keep to normal experience and try to understand that first.

The search we are pursuing seems very broad. For there are challenges and problems faced by all of us today from every background, location and social class. Every day we must all confront issues of health and security, life and mortality. It is partly because of these daily challenges that we might think that religion continues to exist. It is needed by so many. It is the universal tool in claiming to explain things which we do not yet understand scientifically in our essential contacts with the physical world. Or we may do so but we accept that that understanding is only properly held by scientists whilst most of the rest of us are in ignorance of this explanation and understanding. Each of the religions answers questions which humans seem programmed to ask: Why are we here? How shall I live? What happens when I die? And the answers seem to be more directly understandable to most of us than those given by science (where that applies).

How can we answer these deep questions in a manner acceptable to all? I would like to suggest that we each have to discover how to make sense of what goes on inside our minds. Outside is the physical world – which is explained increasingly well by physicists, as I have already emphasized. It is, for each of us the world inside our head which we need to understand. That needs to be achieved in a manner which is understandable to most of us (given we are prepared to work at gaining such understanding).

We should start by investigating that elusive inner sense of self. It has many names, some of which I have already used: inner voice, intuition, inner self, spirit, guide, psyche, 'I' and many more. The best word to describe what is missing, as I have already suggested and used, goes beyond 'I' in order to catch the full richness of our cultural experience. It is our 'soul', a word with many connotations, some to do with experiencing uplift, others mundane. I have already emphasized that it is correctly the soul which gives us the ability to 'be truly there' at the centre of our experience. It is, for each of us, the 'centre point' of our inner Black Hole. It supposedly allows us to know that we are in command of our own mental ship. Through altered states of consciousness – by alcohol, drugs, meditation, prayer or whatever – we can try to enhance or distort our normally fleeting glimpse of this elusive inner sense of 'me-ness'. But we have little knowledge about what it actually consists of. It really is an inner Black Hole, apparently with no content, so it would seem, but in possession of real presence. We might come on it by accident, but then blindly pass it by without ever noticing its existence. So we need to come back to the soul, as the centre both of our understanding and of our complete lack of knowledge. It is the great internal mystery! The great inner Black Hole!

This was truly the case with the influential Scottish philosopher David Hume, who wrote that "Whenever I look inside myself all I can ever catch is a bundle of sensations." So he could find no inner self, no 'I' or soul, however hard he searched.

He had no deeper sense of any internal viewer or owner of his experience. He could only experience the experiences themselves. But then he had no answer to the question "who is the experiencer?" So how could he actually claim there was any experience at all if no-one is having it – quite a paradox, it would seem? He had just about argued himself out of existence! We must try to avoid ourselves getting into that very awkward position.

The search for the meaning of life needs to centre on something other than just the content of experiences – on the bundle of sensations of Hume. We must try to search for our inner self, our soul. But the soul is so elusive and subtle that science has banished it from everyday discussion over the last century and sent it into the outer reaches of deep religious discussion and prayer or to be looked after by New Age fanatics. We already considered earlier the vast swathes of those who should be concerned but dismiss the soul as unworthy to be concerned with or hold the soul so strongly to their breasts that no-one else can easily explore it. That covers the hard-nosed neuroscientists, the genetically-minded psychiatrists, the hollow men and the religious fanatics.

Yet I can only repeat again that the soul is what each of us must possess, in some form or other. It must exist in some Universe otherwise we ourselves are not ourselves – the same old paradox already raised above: an absurdity. I suggest that we must search for this soul, both among the detritus of past meanderings in history, mythology, philosophy, religion and mysticism as well as from advances in psychology, psychobiology and neuroscience, as well as new knowledge being brought to bear on these topics. Only then can we look at the world hopefully in a new light, through the eyes of the soul.

The science of matter has progressed down to making the atom bomb (and killing hundreds of thousands of people by its use) and even to further inside the 'old' elementary particles, down to quarks and gluons and their subtle companions that I mentioned earlier. No similar progress has occurred in the mind. That is why my father could fulminate against such amazing progress in exploring the material world because science has been apparently unsuccessful in explaining what goes on in our own minds. We, and especially our minds, have been left out of the progress leading us to the twenty-first century with its bounty of ipods, ipads and all sort of other electronic gizmos directing us away from our inner selves. But this lag in the understanding of mind was claimed by my father as the clue to why mind is in actuality superior, and will always be less understood, than matter. Mind, to my father, underpins the world in its totality and therefore has to be treated as the senior partner, with the junior one of matter running along behind as a pale copy.

This scenario of the superior mind could be read, for example, into Winston Churchill's pronouncement at Harvard University in 1943

> The empires of the future are the empires of the mind

Indeed that is consistent with the thrust of my book 'The Shape of Minds to Come' which I wrote in 1971 (Taylor 1971). There I had considered how the 'mental revolution' – based on the developing science and related knowledge of the brain and mind – had still 300 years to catch up on the 'matter revolution', started by

Francis Bacon, Galileo Galilei, Isaac Newton and many others. Our understanding of the mind was then, and is still far behind that of matter. However, that disparity is now being removed by the speed of the mental revolution and ultimately they could be neck and neck. Indeed the most recent advances in brain science (for example scanning vegetative states, or detecting what people are actually thinking by just looking at the on-line activity of their brains from the outside of their heads) show that the mind is beginning to give up some of its deeper secrets by its concomitant physical activity of its brain.

So I could not see that 'all is mind' is necessarily the way to look at the relation between matter and mind, especially with the enormous advances being made in brain science. One way to progress is to be prepared to use the scientific method as far as we can. We should take these exciting new results from brain science as part of the 'ground truth' on which to attempt to understand mind. However we need not initially go to the other extreme that 'all is matter', the hard-nosed reductionist approach. We must keep our minds open as to all possibilities. We must add to the initial scientific ground truth whatever else we can find as the facts of the soul.

However, to be able to reach an ultimate conclusion about this subtle and difficult aspect of our being – our soul – our inner Black Hole – we need certainty that we are finally getting somewhere. It is no good spouting off in vague terms about such an important item of our being. It is indeed the very root of our existence. But we must turn to the bedrock of our own experience, properly marshalled and related to the most recent understanding of brain and mind, before we can be assured that the final explanation of our soul and hence a key to the meaning of life is based firmly on all the possible facts and our model for it has been tested to destruction.

Religious experience must be called on to play its role in understanding the mind. It has been and remains such an important determiner of meaning in man's existence that it cannot be shut out. Science must also play its role, as I argued above. It has not helped much so far in understanding the soul, except give us powers over matter far beyond that which we are wise enough to handle (think Hiroshima and Nagasaki). Now we must put science to work, alongside religious experience and any other components seen as useful, finally to uncover the soul. Hopefully it will then give us wisdom about ourselves, allowing us to move forward more securely in the Universe.

At the same time we emphasize again that our attitude to and belief in our own soul's existence must be ground down to the point where only the facts of the case are relevant. Such an approach is difficult due to the weight of cultural bias we all collect as we journey through life. We take on numerous beliefs without knowing or easily being able to acknowledge them. But acknowledge them we must, dragging them into the clear light of day, so that we recognize amongst the bases of our search for the soul which is belief, and which is fact.

As part of this process, I remember being taken some years ago on a tour of the latest big particle accelerator at the Nuclear Physics Centre in Geneva. I was taken over to an enormous hole in the ground going down some 100 m, with a little rope ladder style of device zig-zagging down its side by which we had to descend. I have no head at all for heights and went down gritting my teeth every step of the way. But it was worth it when I got to the bottom – to see the tunnel round which elementary particles were accelerated to enormous speeds very close to that of light then collided with other pieces of matter at one of the spots similar to the bottom of the hole I was in. Above me towered a gigantic measuring system to detect the exotic products of such interactions. If you smash matter up enough then it is broken into all its possible constituents. Digging ever deeper into matter the substructures of substructures of matter are being discovered. These were the new facts being uncovered, to the consternation of some and the delight of others. As I have already remarked, we are now eagerly awaiting results from the next generation particle accelerator – the Large Hadron Collider – to provide us with views of matter never seen before but based on the questions raised by what we already know.

Unlocking the inner secrets of matter is thus proceeding forward, building understanding at one scale or size before moving forward to the next, deeper one. We are now at the scale being explored by the LHC. Undoubtedly we expect that sooner or later we will have to stop building ever bigger machines, since they will become too expensive even for the whole human race's combined financial clout to support. Yet that is a view held and expressed every decade or so over the last century. New ideas to probe the inner nature of matter at an ever smaller scale may well be developed. So the process may go on and on so that we are getting ever closer to the Universe at the very beginning of its Big Bang. That is the optimistic view of our probing the nature of matter. The pessimistic view is one expressed earlier: we will have to stop probing when we run out of money using the present methods. Which wins is actually for the future to decide. But the determiner could well be man's ingenuity. My money is on the continued probing ever deeper.

The overall lesson from this wonderful progress being made in unlocking the secrets of material nature is therefore to attempt a similarly progressive and sure-footed approach in our search for the soul. That is as unknown as is the darker side of matter to be discovered at the Large Hadron Collider. We should be prepared to work as hard at that unknown mental side of our life as the material side that is being explored by the Large Hadron Collider. We possess brain exploration machines of increasingly better sensitivity in terms of the increasingly accurate methods developed over the last few decades to look into the brain of an active, thinking and feeling human. Our brain modeling capabilities are similarly rising to the task as computer power increases, in spite of the subtlety of the brain. So our scientific understanding of brain and mind is catching up with that of matter. Where will it lead us? What may the soul look like under such scrutiny? Will it even be seen by this approach, or will it escape 'under the carpet' so to speak?

1.9 The Soul and the Brain

The science of the brain has been making enormous progress in helping us understand various features of how our minds work and how the brain controls what appears to be all of our behaviour. Only in the last century, however, has it been appreciated just how subtle and complex a system is the brain. It is composed of billions upon billions of very small units called ‘nerve cells’, each cell being a living unit in the brain. It is as if the brain were composed of billions upon billions of little bricks, like a palace. And these bricks can communicate with each other over quite long distances in the brain, far beyond each of their nearest neighbors. The palace of the brain is thereby the most beautiful and complex of any palace in the Universe. Each brain is far more complex than a Black Hole or a Neutron Star, as I have already pointed out, and even more complex than the stars in the heavens. Each of the billions of brains on this earth is packaged so superbly that it can fit inside the average skull, and yet give the owner an impression that he or she has a mind extending out into infinite space. Moreover that mind and brain is special to each person. None of these brains are exactly the same across all seven billion of us here on earth. Only now is the science of the brain able to think it might catch up on the science of matter, in spite of this enormous complexity.

We have begun to comprehend the delicate packaging of the grey matter of the brain in the last hundred years or so. But we have also discovered an inner complexity of each of the brain’s building bricks – the nerve cell – in terms of the complex variety of shapes that each nerve cell could be sculpted into, and the delicate and sensitive way that each cell can communicate with many of its neighbours. Such communication is partly by minute electrical pulses passing between nerve cells, being achieved at the endings of the communication channels (between nerve cells) by chemical means. It is at these chemical endings that drugs can be inserted to change a person’s experience, as in cannabis, crack or heroine usage, or even in the smokers’ nicotine dose that elevates their powers of attention and gives them that necessary ‘lift’. So the chemical sources of experience as well as the electrical ones are becoming exposed to our inquisitive gaze.

The mind is thereby gradually giving up its secrets by the increasingly powerful probing of brain science. As I have already mentioned new and ever more powerful machines have been and are being built to probe brain activity and correlate it with behavior and experience as reported by subjects performing various tasks – trying to detect visual stimuli or sounds, or do other tasks set by experimenters – as they perform them. The brain has increasingly fewer secrets, and even detection of lying or false memory claims can now be probed by brain imaging machines. Yet the problems in front of understanding in detail the brain and its effects on behavior and experiences are still awesome.

In the long run we can rest assured that the work to unravel the brain in action will be done. It is driven on by the needs of all of us to have a better understanding of how to prevent the terrible ravages of mental diseases like Alzheimer’s or Parkinson’s as well as problems presented by depression and schizophrenia, all brain-based. But it is also driven by our need to continue the search – who are we, how do we function, and what is our ultimate mode of being?

My father’s ‘all is mind’ is antithetical to this pursuit of understanding the brain in its details. But the main point of debate between my father and I is as to what is the mechanism which achieves such moving of thoughts around. Is it solely by ‘mental states’ as new and independent entities in the structure of the world, or is it purely the result of material action – by the combined action of those billions of neurons in each of our brains. If we crack that problem then we have begun to make real progress in our search. At least we will know in what sort of Universe it will have to take place. There is also the 'half-way house Universe' in which there is a mixture of mind (as non-physical) and matter (certainly as physical) – a picture subscribed to, I suspect, by many on this planet. So it would help the search immeasurably if the choice between these alternatives could be whittled down. The best way to achieve this would be by finding out the nature of the soul – it must be in one or other of these three possible Universes. But which one?

We have to accept, however, that all that brain imaging and more generally brain science can do is look at the physical activity in the brain. Many would still consider the soul to be too subtle to be seen like that. All the brain activity and related behavior and experience correlated to it have not yet touched the soul, they would claim. It is as if the soul is too subtle to be seen by these coarse brain machines. Nor do we expect to see it by looking down at the level of single nerve cell activity. Indeed at that level – of a single neuron chugging away – it would seem that the soul has disappeared completely. So we could take the attitude that science will be of little ultimate help in our search. The soul just will not appear to our probing into the brain.

To respond to that, let me reply by noting that the soul is indeed elusive, but it must be supported by brain activity of some sort so as to be the foundation of our experience. At least such activity could be probed by new and advanced brain imaging machines. This will give us some important scientific clues as to the nature of the soul. We therefore need to note these aspects carefully and probe them ever more fully. The problem is to think about what such activity would be like – what sort of signature should it have to begin to qualify for its existence that indicates the presence of an inner self? So far there is no clue as to the nature of such activity, which may be one reason (of many) why hard-nosed brain scientists have left the search for the soul well alone.

Without following up the trail of the soul in the manner I have suggested – by attempting to understand all of its physical concomitants – we would not be able to say we have begun to complete the story of 'I'. At the same time we must also follow up on the religious understanding of soul, as well as explore related areas of experience of 'the soul in society'. Thus we have to go down every possible highway and byway of experience in order to ensure that we have properly taken account of the human experience of its soul. All of the mind's experience is grist to the soul's mill. Hopefully hard nosed neuroscientists will soften their noses a little and help pick up the trail.

In spite of such a possibility of a multi-pronged attack on the soul, with science as one avenue of approach, we must accept that science looks an unpromising way of proceeding. As noted above, brain science is moving ahead ever faster. But the mind and especially the soul looks very, very different from the two fistfuls of porridge that make up the brain. The two different elements – mind and brain – have such different characteristics that it would be immediately acceptable to most that the mind can never be properly explained by brain activity. This is especially true for the inner Black Hole of the soul: it can never be created only by two fistfuls of porridge, one could claim. Nor can it be seen by using the telescopes or magnifying glasses of science, whatever they may be. But we will see!

References

Darwin C (1879) The descent of man and selection in relation to sex. John Murray, London
Eliot TS (1922) The waste land. Dial 73(5):473–485. Liveright H (ed) Dial Publishing Company, New York
Sartre J-P (1943) Being and nothingness. Editions Gallimard, Paris
Taylor JG (1971) The shape of minds to come. M Joseph, London
Taylor JG (1973) Black holes: the end of the universe? Ernest Hecht, London

Chapter 2
The Search for Consciousness in the Brain

2.1 Where Is It?

If consciousness is created by brain activity, either solely or in part, then the traces of the relevant brain activity should be able to be observed by suitably subtle experiments and sensitive enough experimental apparatus. Such is the route that has been followed over the last few decades by increasing numbers of neuroscientists to search for what are called the 'neural correlates of consciousness (NCC)' but, however, with rather uncertain results. The main feature of this uncertainty is, I suspect, due to the lack of clarity as to what precisely is to be discovered. In other words there is the difficulty of what exactly the brain activity represents as part of the upcoming conscious experience of a given subject?

It could be the case that the brain supports the total experience itself, as in a purely physical account, or only certain components of consciousness (as in the case of a dualistic model of consciousness, where consciousness is composed of partly physical entities and partly immaterial entities). That there exist purely immaterial entities will be discussed in some detail in a later chapter, where it will be strongly argued that such entities do not exist, at least as part of a Universe of discourse in which all entities can be considered in a scientific manner, able to be probed to as great a depth as needed for their exploration.

Given, then, that we are left only with physical entities with which to work, that leaves us, in the case of the brain, with its basic living entities – the nerve cells or neurons – of which the brain is so constituted. It will be out of the interactions of these somewhat complex living units that we must attempt to construct the elements of consciousness in a manner able to be tested by scientific experimentation.

The brain, then – from this point of view – is the main physical entity with which we have to deal to understand consciousness. But the brain is an awesomely difficult, complex system with which to deal, as noted before. It is composed, as already mentioned in the previous chapter, of the order of 100 billion nerve cells,

J.G. Taylor, *Solving the Mind-Body Problem by the CODAM Neural Model of Consciousness?*, Springer Series in Cognitive and Neural Systems 9,
DOI 10.1007/978-94-007-7645-6_2, © Springer Science+Business Media Dordrecht 2013

with these occurring in a variety of different types (of the order of 20–30) with different dynamical responses to inputs and with differences arising from their possible different places in the brain.

Over the centuries the brain has begun to give up its secrets, first in terms of the clumps of nerve cells of which it is composed and then in terms of the more detailed microstructures by means of which the nerve cells are interconnected. The functions of the different so called ‘grey matter’ nuclei or clumps of nerve cells into which the brain may be divided, has been gradually teased out of the brains of subjects who have lost the related nuclei, through disease (such as stroke) or from accidents, such as occur from car accidents or from injuries sustained in wars. More recently functional magnetic resonance imaging (fMRI) has been used to give ever greater precision as to what these various regions in the brain are doing – what is their functionality – in normal living brains.

In particular the division of the brain, first into cortical and sub-cortical regions and then into sub-divisions of the nerve cells of the cortex into suitably similar regions, has proceeded apace. The cortex itself, the grey ‘rind’ on the surface of the cerebral hemispheres, has been divided up in this manner into numerous areas, the most popular now following the original division into the Brodmann areas. But even before that it was recognized that there were several ‘lobes’ in the cortex, with different functions. These are:

- The visual cortex (occipital lobes), devoted to analysis of the visual stimuli entering from the eyes;
- The temporal lobes, involved in coding for objects as seen in the environment;
- The parietal lobes, heavily involved in motor actions, such as grasping objects, as well as in short term memory;
- The prefrontal lobes, concerned with higher cognitive processes such as setting and attempting goals and making decision between different choices.

Whilst there is much support for the cortices as being the site of the creation of consciousness, other sub-cortical sites have been proposed as either relevant or even the site of such creation. Thus the brain stem and related area have been proposed by

Damasio (2000) as the site, especially for the development of early consciousness (termed by Damasio ‘proto-consciousness’: Damasio 2000). Sites in the brain stem, such as the periaqueductal gray (PAG), have been proposed by Panksepp as being crucial for consciousness (Panksepp 2005). I proposed the nucleus reticularis thalami, or NRT, as a crucial site in the development of consciousness, being an important region for allowing a competition between activities in different cortical sites, possibly to gain consciousness (Taylor and Alavi 1995). From the great degree of connectivity of a particular subcortical site, the thalamus was proposed (Crick 1994) as the main site in the brain supporting consciousness creation.

However the main thrust on the siting of consciousness in the brain, or at least an important component of the total brain network of modules needed for consciousness to arise of a given stimulus, is that of various modules (areas of brain tissue) in the cortex. Which lobes are important has been much debated, such as the temporal lobe, the occipital lobes or visual cortices, the parietal lobes

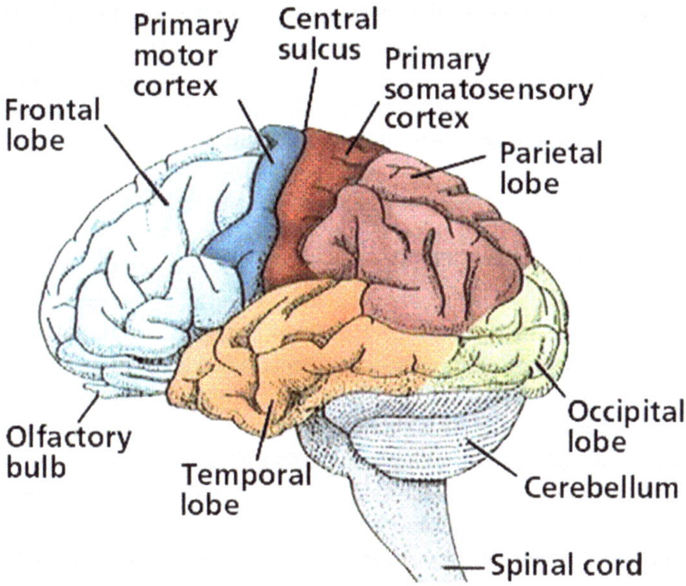

Fig. 2.1 The various lobes of the brain

or the prefrontal cortices. These various lobes of the brain are shown in Fig. 2.1. All of these regions may indeed contribute to the emergence of consciousness itself; as we shall see later we expect these various components will be involved in giving meaning to the object representations at the basis of our consciousness. Thus besides the temporal lobes for such object representation, there will also be needed prefrontal cortical activity to represent goals to be achieved, visual cortical activity to allow the details of the object representations to be fleshed out, and parietal lobe representations to give both spatial structure and place to the object representations as well as possible manipulations that can be carried out on the objects (already learnt by experience).

Of course it may be necessary to give more detailed specification to the regions of greatest concern in the emergence of consciousness of an external stimulus. However we will not address that in greater detail till later in the book. For the present we should accept that such emergence is caused by some very complicated dynamics between the activity in various brain regions, most likely distributed throughout the brain.

2.2 The Neural Networks of the Brain

It is necessary, for completeness in the book, to introduce briefly some of the main ideas of neural networks. For, as we have emphasized above, the creation of consciousness will most likely occur through some form of complex dynamical

2.2.2 A Simple Mathematical Model of the Neural Networks of the Brain

We might hope, in our search for consciousness, that a suitably powerful mathematical framework for the brain would somehow lead to our goal. We might further hope that the mathematical structure, if rich enough and in general close enough to mimic the structure of the brain, would lead us to an obviously conscious system. Let us try.

Along these lines, then, there has been developed the mathematical framework of what is called 'continuum neural field theory' as the beginning point for an analysis of the multi-modular global brain and its possible consciousness. This approach assumes that the cortex is composed of a continuum of nerve cells, so is to be seen as a sheet of such cells. Arguments for and against this approach are immediate, but the continuum model is expected to be a good approximation to the actual living sheet of nerve cells in different cortical areas, as observed at a suitably large scale. We take this approach since it is elegant, leads to a reasonably good approximation to that of the brain, and has mathematical features amenable to direct mathematical analysis (which can be used to explain various features of brain activity and of experience).

The brain is an amazingly complex dynamical system. This has been found to be true at all levels of its investigation: from overall dynamics of the global brain down to the functioning of synapses, in which a wealth of different neuro-chemicals are involved. We can usefully contrast those studies at the micro-processing level of the brain (at the synapses and ion channel level) in comparison to studies at the global level The former are involved with problems containing systems of variables more controllable than in the global case, as are the possibilities of performing experiments to test predictions of models. That is even possible at a still higher level, as for example in recent careful 'bottom-up' models of the cerebellum in its role in conditioned learning (Ohyama et al. 2003). However there are only a few detailed models of the dynamical interactions of modules across the vast reaches of the brain. Even these tend to be based on simulations, and not on any mathematical principles (although there are mathematical equations behind such simulations, even if these are ad hoc). Instead of analysis at the level of micro-structure of nerve cells and of their connections, it is more appropriate in this book to try to develop modeling and analysis of the global brain. This will be in a manner that emphasizes the functions being performed by the various modules (Brodmann areas, for example) using rather a simplified view of each nerve cell, and even of the overall global brain itself. We do that in terms of recent functional proposals for overall control systems in the brain: of attention (Corbetta and Shulman 2002; Taylor and Fragopanagos 2005), of motor control (Miall and Wolpert 1996; Wolpert et al. 1998), of emotions (Pessoa et al. 2002; Taylor et al. 2003), and even of consciousness (Taylor 2003).

This program is started in the next sub-section, where the basic components of the approach will be presented. This is then formulated in detail mathematically in the Appendix, where expected results from the mathematical analysis are developed. A control version is also developed there (for those interested to see how such mathematical analysis might proceed). A brief summary of results is then given, especially as to the possibility of the emergence of some form of consciousness in the ensuing dynamics.

2.2.3 *The Brain's Basic Components*

We have already discussed the main problem we face in any program to construct a mathematical theory of the global brain: the brain possesses too much complexity. To underline this point, we have to face the hierarchy composed of:

- Chemical components of synapses
- Channel variable dynamics for each synapse
- Overall dynamics of each neuron compartment
- Overall dynamics of each neuron
- Overall dynamics of each column
- Overall dynamics of each module
- Overall dynamics of the global brain.

If each level were attempted to be modeled faithfully we would have enormous complexity of the system by the time we arrive at describing the overall brain itself. One way to proceed is by making a sequence of ever more complex approximations. The starting point would then reduce to a problem which looks as if it has a chance of being solved reasonably well. Further additions of complexity, going back up the bullet points above, would regain the brain in all its complexity, in the process having some idea of what the principled powers were of the basic 'first approximation', and what further powers are needed, and might be added, by adding the further bullet-pointed properties.

The first step – of formulating the basic or simplest approximation – is very important. It must be so structured as to lead us to expect interesting features that would be able to lead to some of the true powers of the brain in reality. But it must be simple enough to be relatively soluble, by which I mean allowing the deduction of existence theorems and characteristics of some of the solutions, but not necessarily providing analytic forms of solutions.

As suggested earlier in this chapter, a particular simple and analyzable first approximation which has drawn considerable interest, but only at the single (or few) module levels is that of continuum neural field theory (CNFT). One pioneer of this was Amari (1977) who proved in 1977 some remarkable features of a certain class of CNFT models in 1-dimension (so composed of lines of interconnect nerve cells): the existence of long-term localised solutions, or 'bubbles' of restricted neural activity. This was extended more recently to the 2-dimensional case (Taylor 2000a, b, c). A little later Amari and colleagues developed a mathematical analysis of learning of afferents in the CNFT framework, in terms of the distribution of inputs (Takeuchi and Amari 1999). That has since been applied to the brain in a variety of ways: showing the development of known orientation selectivity in a

(d) A summary of these and related experiments was given by Driver and Mattingley (1998) as “The loss of awareness in neglect implies that the inferior parietal lobe and its connections contribute to visual awareness.”

2.3.2 *Where Is Consciousness in the CR?*

Discussions on the siting of the various higher level processing involved in the CR can proceed apace, but do not seem to lead to an understanding of the manner in which the CR can contribute to our understanding of consciousness. However in (Taylor 1999) a crucial new feature was introduced into the CR: that of attention. Indeed figure 5 in that reference provided a model of attention as a control system, including a corollary discharge system. This was supposed to provide initial inhibition to possible distracters entering the relevant working memory buffer, so leaving the buffer clear for the triumphant entry of the neural activity representing an attended object in the environment. It was this process of inhibition, by means of a corollary discharge of the attention movement control signal, which was the important clue to how inner consciousness could arise, as will be developed more fully later.

At the time this extra corollary discharge of attention component was brought in there was no experimental support for it nor any interpretation of it in terms of any understanding of consciousness that was available at that time (Taylor 1999) However I persisted in developing the model of attention control, to be discussed in more detail throughout this book, and in fact presented it at the gargantuan Society for Neuroscience meeting in 2000 (Taylor 2000a, b, c). No notice was taken of it, and a decidedly hostile group was developing its weapons within the next few years to scupper any idea that attention and consciousness were connected (as will be discussed in more detail in the next chapter).

This was and is an important point to clarify: does the consciousness of a stimulus require attention being paid to it or not? That is a question that has dogged the subject of consciousness since then and needs to be addressed in full.

2.4 So Where Is Consciousness in the Brain?

We have so far attempted to answer this question by means of a two-pronged attack. One is through the development of a Continuum Field Theory (CNFT) model of the overall brain, with its bubbles created by the dynamics of the continuum, and as presented in the Appendix. However this was only achieved in a very schematic manner, without enough detail to make it convincing as to its relevance to consciousness creation. More detail perhaps needs to be included in the CNFT model, such as making it a spiking CNFT (SCNFT) or a spiking stochastic CNFT

(SSCNFT). Yet the dynamics of the CNFT, the SCNFT or the SSCNFT will still only appear to be ever more complex and noisy versions of the bubble dynamics we have already explored rather briefly. Such models could also support oscillatory dynamics, so altogether fitting much of the dynamical behavior already observed in the living brain.

The difficulty we face is that this activity gives us no hint of any emergence of consciousness in the ensuing 'experience' of the owner of that brain. We still have no clue as to how such experience, of some sort of 'inner self', could arise as part of these dynamical features of the brain activity. How can we proceed further?

We have already included in the CNFT model of Sect. 2.2 and the Appendix an attention control system (especially in the appendix). We would have hoped that a suitably clear and precise model of attention would require or produce, somewhere in its interstices, exactly the signal of consciousness' inner experience that we have been hoping for. That has not happened.

At the same time we need to take account of a strong movement of activity attempting to show that attention and consciousness are not so incredibly interfused as we have been assuming. Perhaps these other ideas may lead us into a completely new approach to consciousness, untrammeled by the strains of having attention inalienably fused with consciousness in some as yet unknown manner. We turn to this escape route next.

Appendix: A Continuum Neural Field Model of the Brain

A.1 Mathematics of the Simple Brain

A CNFT model is based on the approximation of a module of neurons as composed of a two-dimensional sheet of neurons, with the neuron at the two-dimensional position **r** on the sheet having membrane potential V(**r**). The simplification in this 'sheet-like' assumption allows us to consider many neurons at once, although we can relatively easily reduce the sheet back to a finite number of neurons by using the localised distributions of neurons at a finite set of points on the sheet.

The dynamics of the neurons is also greatly simplified by assuming a graded response pattern for each neuron output (although this can be extended to spiking neurons if needed). Thus the output of each neuron is taken to be some function f( ) of its potential; the simplest case being a step function, although results have been obtained for more general sigmoid functions. The simplest dynamics is taken as

$$\tau dV(\mathbf{r})/dt = -V(\mathbf{r}) + w^{*}f(V)(\mathbf{r}) + I(\mathbf{r}) - h \tag{2.1}$$

where I is the external input to the module at that point, w(**r, r′**) denotes the lateral connection strength between the neuron at **r′** and that at **r**, −h is a constant inhibitory bias to all neurons to assure stability and suitable competition between neurons, and * is the usual symbol for the convolution product taken over the positions of the module, defined as:

$$\mathrm{w}^{*}\mathrm{f}\,(\mathrm{V})\,(\mathbf{r}) = \int \mathrm{w}\left(\mathbf{r}, \mathbf{r}'\right) \mathrm{f}\left(\mathrm{V}\left(\mathbf{r}'\right)\right) \mathrm{d}\mathbf{r}'$$

We now extend Eq. (2.1) to a set of interacting modules as

$$\mathbf{T}\mathrm{d}\mathbf{V}\,(\mathbf{r})\,/\mathrm{dt} = -\mathbf{V}\,(\mathbf{r}) + \mathrm{W}^{*}\mathrm{f}\,(\mathbf{V})\,(\mathbf{r}) + \mathbf{I}\,(\mathbf{r}) - \mathrm{H} \tag{2.2}$$

where the extension of (2.1)–(2.2) is achieved by taking **V** to be a vector-valued field of membrane potentials (each component denoting the neural field for a given module), W(**r**′, **r**) now denoting the matrix of field connections, with diagonal connections being the lateral one entering originally in (2.1), the off-diagonal ones being those connecting different modules, −H is now a diagonal matrix, with constant values in each entry for a given module although with possible differences across modules to allow different levels of overall inhibition, and **I** denoting a vector field of external inputs, each component again being associated with a given neural field module; the matrix **T** denotes a diagonal matrix of time constants (where also in (2.1) this can be extended to different time constants for different neuron positions if so desired). We note that we are simplifying by taking the same co-ordinates for each module; again that can be generalized. Further we are only taking neurons of the same type in (2.1) and (2.2); again that can be extended to those of inhibitory and excitatory form or of different sub-populations, by suitably extending the notation in these equations.

We now indicate some of the features expected from (2.2), extending those possessed by (2.1) – bubble existence, dynamics of bubbles and learning structures.

(a) Basic Features of the Dynamics
A Liapunov function can be derived for the dynamics of (2.2) extending that for (2.1) in the case of symmetric connection matrix W (across modules as well across the lateral connections in each module), as in (Giese 1998), thus providing general stability arguments.

(b) Existence of Bubbles
The one- and two-dimensional bubble analyses of (Amari 1977; Taylor 2000a, b, c), and the many simulations in the references (Fellenz and Taylor 2000; Taylor 1997; Petersen and Taylor 1996; Petersen 1997; Trappenberg et al. 2001; Giese 1998; Stringer et al. 2003a, b), lead to the expectation that there will exist multiple bubbles, across a range of modules. It is possible to develop equations for coupled bubbles from (2.2).

(c) Dynamics of Bubbles
Bubbles have been found (Amari 1977; Taylor 1997, 2000a, b, c; Takeuchi and Amari 1999; Fellenz and Taylor 2000; Petersen and Taylor 1996; Petersen 1997; Trappenberg et al. 2001; Giese 1998; Stringer et al. 2003a, b) to be driven to

flow to regions of highest input. A similar situation is expected to occur for the coupled bubbles in the expanded version (2) of CNFT.

(d) Learning Structures
The original one-dimensional work of (Takeuchi and Amari 1999) was extended in (Taylor 2000a, b, c) to two dimensions and to applications to specific brain modules in (Fellenz and Taylor 2000; Taylor 1997; Petersen and Taylor 1996; Petersen 1997; Trappenberg et al. 2001; Giese 1998; Stringer et al. 2003a, b). The most crucial feature of this study was the presence and exploitation of instability in the learning law dynamics, producing discontinuous periodic structures mapping higher dimension input spaces down to the two-dimensional sheet in a clearly defined, even analytic, manner. Similar structures are to be expected in the extended case of (2.2).

It might be supposed that the system of Eq. (2.2) can be reduced to an equation identical to (2.1), though now with the position vectors of the neurons as obtained, for example, by squashing the various modules together in some way. The single lateral interaction weight function W(r, r′) (with r and r′ running over the whole set of adjoined modules) will now be highly non-local, with no structure such as a Mexican hat or other locally bounded function which is centrally positive and then turns negative far enough away from the origin (as arises in each component of the lateral connection matrix W in (2.2)). Each component of the matrix W introduced in (2.2) will have now lost its intuitive character in the new squashed representation with the form of (2.1).

It is clear that the module-based structure of (2.2), containing as it does an inherent similarity to the cortex, is of great value to obtain extension to the hierarchy of modules of the results such as those obtained in (Amari 1977; Taylor 1997, 2000a, b, c; Takeuchi and Amari 1999; Fellenz and Taylor 2000; Petersen and Taylor 1996; Petersen 1997; Trappenberg et al. 2001; Giese 1998; Stringer et al. 2003a, b), as any summary reading of these papers would show. This is especially so for those results which depend crucially on the Mexican hat structures used in detailed simulation of psychophysical results (Fellenz and Taylor 2000; Taylor 1997; Petersen and Taylor 1996; Petersen 1997; Trappenberg et al. 2001; Giese 1998; Stringer et al. 2003a, b). In the new ‘squashed’ framework, however, we lose the transparency needed to extend the various results of (Amari 1977; Taylor 1997, 2000a, b, c; Takeuchi and Amari 1999; Fellenz and Taylor 2000; Petersen and Taylor 1996; Petersen 1997; Trappenberg et al. 2001; Giese 1998; Stringer et al. 2003a, b) from a single module to a hierarchy of such modules. Thus we would not expect it will be easy to prove, in the squashed representation, the results of, say, the existence of locally bounded ‘bubbles’ (Amari 1977; Taylor 2000a, b, c; Trappenberg et al. 2001; Giese 1998; Stringer et al. 2003a, b) coupled across various modules, although in general with different sizes, or of the development, through Hebbian learning, of suitably periodic structures of afferent weights from one module to the next in the original hierarchy (as known experimentally, although with increasing wavelength in going from V1 to V2 to V4 and so on (DeYoe and Van Essen 1988), especially fig 24.6 in that reference). Nor will the detailed properties of these solutions be easily derivable. This is due to the loss of the ability to extend all of the elegant properties of solutions of (2.1) upwards in the hierarchy by means of the transparency of the structure of (2.2) when working in the far less intuitively understandable squashed multi-modular system being considered as an alternative.

The brain has not used such a 'squashed' representation to perform its dynamics, in that the separate modules are kept geographically separate. It would appear also to be computationally inefficient to have such a representation, or nature would have preferred it and squashed the modules all together.

A final general point about the structures of (2.1) and (2.2) is concerned with the inhibitory bias, which arises as the inhibitory constant fields arising from the constant h in (2.1) or its extension to the diagonal matrix H in (2.2). These have a strong effect, as any reading of the papers of (Amari 1977; Taylor 1997, 2000a, b, c; Takeuchi and Amari 1999; Fellenz and Taylor 2000; Petersen and Taylor 1996; Petersen 1997; Trappenberg et al. 2001; Giese 1998; Stringer et al. 2003a, b) indicate, in association with the existence of bubble solutions and periodic unstable afferent connection strength learning equations. The constant(s) could be absorbed into the definition of the potential V(r), but again this hides the meaning ascribed to V as corresponding to a local, laterally connected field, vanishing at infinity. This is all part of the general structure of the CNFT approach, as more fully described in (Amari 1977; Taylor 1997, 2000a, b, c; Takeuchi and Amari 1999; Fellenz and Taylor 2000; Petersen and Taylor 1996; Petersen 1997; Trappenberg et al. 2001; Giese 1998; Stringer et al. 2003a, b).

Let us consider how the structure given by (2.2) can be used to extend some of the specific results of (2.1). We consider two modules only, as has already been considered in (Taylor 1997) in the case of color illusions, involving two color modules, one for each color, and the bubble solutions found to fit the patterns observed when a blue/red boundary was stabilized on subjects' retinas (one being the non-intuitive 'sea' of blue and red mixed together, as experienced by some subjects). In the asymptotic limit in time, (2.2) becomes:

$$u(x) = \int W(x - x') f[u(x')] dx' + \int W'(x - y') f[v(y')] dy' + h \quad (2.3a)$$

$$v(y) = \int W''(y - y') f[v(y')] dy' + \int W'''(y - x') f[u(x')] dx' + h' \quad (2.3b)$$

where we have assumed translation invariance of the interconnection matrices, and the variables x, etc. denote suitable co-ordinates on the cortical surface. In terms of the standard notation (Amari 1977; Taylor 2000a, b, c), and defining the function

$$U(x) = \int_0^x W(y) dy \quad (2.4)$$

we can rewrite (2.3a, b), for a local 'double bubble' solution, with x restricted to the interval [a, b] and y in the interval [c, d], as

$$\begin{aligned} u(x) &= [U(x-a) - U(x-b)] + [U'(x-c) - U'(x-d)] + h \\ v(y) &= [U''(y-a) - U''(y-b)] + [U'''(y-c) - U'''(y-d)] + h' \end{aligned} \tag{2.5}$$

where

$$u(a) = u(b) = v(c) = v(d) = 0 \tag{2.6}$$

Applying the constraints (2.6) to (2.5), we obtain an extension of the set of Eq. (2.6) of (Amari 1977), as

$$\begin{aligned} 0 &= h + U(b-a) + [U'(a-c) - U'(a-d)] \\ 0 &= h + V(b-a) + [U'(b-c) - U'(b-d)] \\ 0 &= h' + U'''(d-c) + [U''(c-a) - U''(c-b)] \\ 0 &= h' + U'''(d-c) + [U''(c-a) - U''(c-b)] \end{aligned} \tag{2.7}$$

The Eq. (2.7) can be investigated for coupled 'bubble' solutions, and these can be found, as can extension of the theorems of (Amari 1977) on the nature of these solutions (as also discussed in (Taylor 1997)).

The system of Eq. (2.5), and the time dependent form of equations they arose from in (2.2), can also be analyzed for stability, by using a first-order perturbation approach. The basic result is the coupled equations (in a set of coupled module indexed by i, j)

$$\begin{aligned} d\Delta_i/dt = (1/\tau_i)\Big[(1/c_{i1} + 1/c_{i2}) \Big\{ \big[\Sigma_j W_{ij}\left(x_{i2} - x_{j1}\right) - W_{ij}\left(x_{i2} - x_{j2}\right)\big] \\ + h_i - \big[\Sigma_j W_{ij}\left(x_{i1} - x_{j1}\right) - W_{ij}\left(x_{i1} - x_{j2}\right)\big] \Big\} \Big] \end{aligned} \tag{2.8}$$

where $\Delta_i = x_{i2} - x_{j1}$ is the width of the bubble in the i'th module, and with the definition $c_{i1} = \delta u_i\ (x_{i1},\ t)/\delta t$, and similarly for c_{i2}. These equations extend those of stability in (Amari 1977), and so allow for the determination of the stability of bubbles and for their movement in the presence of new inputs, in the coupled module situation. Again these results could not easily have been obtained by using the squashed version of the coupled module Eq. (2.2).

A.2 Insertion of Control Structures

So far we have only extended the standard CNFT model of a single module to that of several such modules, without any understanding of how functional differentiation can be included in the system. We now turn to that important aspect.

To justify our approach we need to accept that we cannot expect our extended model to learn its feed-forward and feedback connections all on its own, without any use of genetic memory. This was built up over hundreds of millions of years by pressure of the environment, which is changing all the time. Such pressure has led to crucial functional variations between modules, that allow them to be functionally differentiated into input processing modules, semantic map modules, higher level control modules and response modules, as well as others (and also with differentiation at sub-cortical level).

The first and fourth of these types of modules have already been discussed in the brain context in (Fellenz and Taylor 2000; Taylor 1997; Petersen and Taylor 1996; Petersen 1997; Trappenberg et al. 2001; Giese 1998; Stringer et al. 2003a, b). Here we turn specifically to the third class of modules, those for attention control, by suitable assumptions on the lateral connection matrix W in terms of the depth, strength and width of the lateral connection matrix internal to a module, as well as by temporal flow of activity. The connection matrix elements affect the size of bubbles, and the overall level of the WTA nature of the module. The higher level modules in parietal lobe will therefore be allocated large values of inhibitory connections so as to provide a strong bias towards competition and hence generation of attention control signals.

The temporal flow of activity of the brain has been observed in many ERP studies. It is observed that early input flows through low-level sensory cortices rapidly to prefrontal cortex, and the incoming information is then used to control later processing by feedback through parietal and temporal lobes. Such a flow pattern impresses on the brain a clear functional differentiation: prefrontal cortices act as goal systems to control more detailed lower level processing.

A set of coupled CNFT equations were developed in (Taylor and Rogers 2002), and used as the basis of simulation of the Posner attention benefit effect, based on the above features. The simulation used differentiation of function both by differences in lateral connectivity and inhibition across modules as well as by differences in the temporal flow of activity across modules. In particular we incorporated in (Taylor and Rogers 2002) the early flow of activity to frontal lobes so as to act as an exogenous goal bias to the attention movement controller in parietal lobe, and thence feeding back activity to the sensory (and motor) cortices. This feedback is taken to be achieved by attention contrast gain, so is applied in a quadratic sigma-pi manner to the input on the sensory cortical neurons. In terms of three modules, the goal module with neurons with activity g(r), the attention movement controller with activity v(r) and the input sensory module with activity u(r), the resulting control equations are (Taylor and Rogers 2002):

$$\tau du/dt = -u + w^{*}I + w'^{**}I \times f(v) - h \tag{2.9}$$

$$\tau dv/dt = -v + w''^{*}f(g) \tag{2.10}$$

$$g = g_{des}$$

where g_{des} is the desired goal, and the dynamics in the goal system is being neglected. We can include both endogenous and exogenous attention goals by choosing g_{des} as the external input (in the exogenous case) or as a given externally determined activation to the goal system in prefrontal cortex. In (2.9) we use the notation $w^{**}I \times f(v)$ to denote the quadratic sigma-pi contrast gain amplification input, with

$$w^{**}I \times f(v) = \int w(r, r', r'')\, I(r')\, f(v(r''))\, dr' dr'' \tag{2.11}$$

as was used in [21] and references therein.

We have already noted that the attention control system (2.9), (2.10) can be used to simulate the benefit in response time to an attended stimulus and to related competitive processes. It has been extended to sensory motor attention control (Taylor and Fragopanagos 2005; Taylor 2003).

More generally, the above control model has been extended so as to contain a monitor and a predictor of the future state, as in the CODAM model (Taylor 2003) and applied recently to simulate the attention blink in (Fragopanagos et al. 2005). This was achieved by including a working memory or buffer module WM and a monitor MON. The new system of equations which update the set (2.9), (2.10) are now

$$\tau du/dt = -u + w^{*}I + w'^{**}I \times f(v) - h \tag{2.12}$$

$$\tau dv/dt = -v + w''^{*}f(g) + w'''^{*}F(MON) - w''''^{*}f(WM) \tag{2.13}$$

$$g = g_{des} \tag{2.14}$$

$$MON = |WM - g| \tag{2.15}$$

$$\tau' dWM/dt = -WM + f(u) \tag{2.16}$$

where we have assumed that the input to the WM from the input processing layer is mainly excitatory. Moreover we have included a different time constant for the working memory buffer WM in (2.16) in order to allow for longer time constants (although we used in the AB simulation in (Fragopanagos et al. 2005) a set of reciprocally-coupled neurons as having greater flexibility, rather than manipulating the neuron time constants).We also use the extended Hebbian learning law

$$\delta w'(x, x', x'') = \varepsilon f(u(x)).f\Big(v(x'').I(x')$$

where ε is a learning rate.

In Eq. (2.13) (and developed more fully in (Fragopanagos et al. 2005)) we included inhibitory feedback from the WM system to all nodes not coding for the input I in the attention movement In this way attention becomes a scarce resource: if a large attention load must be processed, say with many distracters, then the error can become large in the MON module, so will boost the effect of creating the attention control signal, as in (2.13).

We still have to face hard learning problems. The possibility of applying developmental knowledge to the learning process also needs to be considered. This can be achieved by including learning in an incremental fashion, so that lower level representations will be learnt first and stabilized before further learning, under top-down control, is allowed. Furthermore the manner in which goal representations in prefrontal cortex could arise needs to be considered.

Emotion is minimally included by the addition of 'valence' modules (amygdala and orbito-frontal cortex), following the emotion brain architecture already presented elsewhere (Taylor et al. 2003), but now represented in the CNFT framework. We refer the interested reader to these developments, and to the results of Chap. 13 for a more complete discussion.

A.3 Results of the Program

The basic results of the above program are of three sorts:

1. Understanding of the level of coupled bubble formation and dynamics, under simple feed-forward- & feedback coupling assumptions, with sizes and expected influences of bubbles on each other determined by relative parameter choices and fan-in values in the various modules. Some progress on this has already been made (Taylor 1997), and further structural results were presented in Eqs. (2.3), (2.4), (2.5), (2.6), (2.7), (2.8), (2.9), (2.10), and (2.11);
2. Learning of cortical representations, both of feedback and feed-forward form can be obtained, supporting topographic spatial and localized object representations (using pre-specified fan-ins depending on the site of the CNFT module being considered.
3. Provision of a basis for addition of further complexity into the system, as well as applying other criteria, such as information maximization, to constrain the approach. One of the unknowns in a general control problem is as to the quantity (if any) being optimized in the control system; for the brain it is expected to be a function of the total reward, although this cannot be evaluated solely by the net dopamine influx (since there are internal sources).
4. It could be argued that the bubbles of neural activity arising in certain sites in the above coupled modules of the CNFT approach could be interpreted as generating consciousness. But that is an unsupported conjecture, needing much more careful discussion of the nature of these bubbles. There are also many other features left out of such an identification: the relation to attention, the fact that prolonged but

non-conscious brain activity can be detected in various sites in the brain. But the main problem is that there is rather little detailed analysis of the various modules in the coupled CNFT approach. Such analysis is needed to relate to the many results from brain imaging experiments on various aspects of consciousness.

A.4 Conclusions

The above discussion provides a general framework with which to attack the dynamics of the brain. It allows stable state analysis as well as extension to the temporal dynamics of a set of interacting CNFT modules. Learning presents also dynamical features that allow the analysis of pattern structure of the synaptic weights. The nature of emotional modulation has yet to be properly inserted by use of reward learning, but this can be included in subsequent versions of the multi-modular CNFT brain model. Much work lies ahead, but general features have already been obtained that indicate the value of the approach.

Open questions still to be faced in the above framework are numerous, but among them can be singled out:

(a) What is the manner in which synchronization of neural activity is achieved (by extension to spiking neurons)?
(b) What is the mode of operation of acetylcholine as a modulator both of the neural dynamics and the synaptic learning process so as to arrive at the contrast gain attention amplification feedback of Eqs. (2.8) and (2.10)?
(c) What level of interaction occurs between modalities and between sensory processing and motor response (as observed as in the psychological refractory effect)?
(d) How can the more complex mathematical structures obtained by adding to the above CNFT model the increasing complexity of the real brain, (as indicated in Sect. 2.2), be analyzed?
(e) How important is the greater complexity granted to the living brain, and which particular components are most important?

References

Amari S-I (1977) Dynamics of pattern formation in lateral-inhibition type neural fields. Biol Cybern 27:77–87

Awh E, Smith EE, Jonides J (1995) Human rehearsal processes and the frontal lobes: PET evidence. Ann N Y Acad Sci 769:97–117

Corbetta M, Shulman GL (2002) Control of goal-directed and stimulus-driven attention in the brain. Nat Rev Neurosci 3:201–215

Crick F (1994) The astonishing hypothesis. Scribner, New York

Crick F, Koch C (1990) Towards a neurobiological theory of consciousness. Semin Neurosci 2:263–275
Damasio A (2000) The feeling of what happens: body and emotion in the making of consciousness. Mariner Books, New York
Dehaene S, Naccache L, Le Clec HG et al (1998) Imaging unconscious semantic priming. Nature 395(6702):597–600
DeYoe EA, Van Essen DC (1988) Concurrent processing streams in monkey visual cortex. TINS 11(5):219–226
Driver J, Mattingley JB (1998) Parietal neglect and visual awareness. Nat Neurosci 1(1):17–22
Fellenz W, Taylor JG (2000) Establishing retinotopy by lateral-inhibition-type homogeneous neural fields. Neurocomputing 48:313–322
Fragopanagos N, Kockelkoren S, Taylor JG (2005) A neurodynamic model of the attentional blink. Brain Res Cogn Brain Res 24(3):568–586
Giese MA (1998) Dynamic neural field theory for motion perception. Kluwer Academic Publishers, Boston
Miall RC, Wolpert D (1996) Forward models for physiological motor control. Neural Netw 9:1265–1279
Ohyama T, Nores WL, Murphy M, Mauk MD (2003) What the cerebellum computes. Trend Neurosci 26:222–227
Panksepp J (2005) Affective consciousness: core emotional feelings in animals and humans. Conscious Cogn 14:30–80
Pessoa L, McKenna M, Gutierrez E, Ungerleider LG (2002) Neural processing of emotional faces requires attention. Proc Natl Acad Sci U S A 99(17):11458–11463
Petersen R (1997) Modelling learning in the brain. Ph.D. thesis, University of London, (unpublished)
Petersen RS, Taylor JG (1996) Reorganization of somato-sensory cortex after tactile training. In: Touretsky DS, Mozer MC, Hasselmo ME (eds) Advances in neural information processing systems. MIT Press, Cambridge, MA, pp 82–88
Stringer SM, Rolls ET, Trappenberg TP, de Araujo IET (2003a) Self-organising continuous attractor networks and motor function. Neural Netw 16:161–182
Stringer SM, Rollls ET, Trappenberg TP (2003b) Self-organizing continuous attractor networks with multiple active packets, and the representation of space. Neural Netw 17(1):5–27
Takeuchi A, Amari S-I (1999) Formation of topographic maps and columnar microstructures in nerve fields. Biol Cybern 35(2):63–72
Taylor JG (1997) Perception by neural networks. Neural Netw World 7:363–395
Taylor JG (1999) Towards the networks of the brain: from brain imaging to consciousness. Neural Netw 12:943–959
Taylor JG (2000a) The central representation: the where, how and what of consciousness. In: White KE (ed) The emergence of mind. Fondazione Carlo Erba, Milan, pp 117–148
Taylor JG (2000b) A control model for attention and consciousness. Soc Neurosci Abstr, 26, 2231#839.3
Taylor JG (2000c) Neural 'bubble' dynamics in two dimensions: foundations. Biol Cybern 80:393–409
Taylor JG (2001) The central role of the parietal lobes in consciousness. Conscious Cogn 10:379–417
Taylor JG (2003) Paying attention to consciousness. Prog Neurobiol 41:305–335
Taylor JG, Alavi FN (1995) A global competitive neural network. Biol Cybern 72:233–248
Taylor JG, Fragopanagos N (2005) The interaction of attention and emotion. Neural Netw 18(4):353–369
Taylor JG, Rogers M (2002) A control model of the movement of attention. Neural Netw 15:309–326

Taylor JG, Fragopanagos N, Cowie R, Douglas-Cowie E, Fotinea S-E, Kollias S (2003) An emotional recognition architecture based on human brain structure. In: Proceedings of ICANN/ICONIP 2003. Springer, Berlin, pp 1133–1140

Trappenberg TP, Dorris M, Klein RM, Munroe DP (2001) A model of saccade initiation based on the competitive integration of exogenous and endogenous signals from the superior colliculus. J Cognit Neurosci 13:256–271

Wolpert DM, Miall RC, Kawato M (1998) Internal models in the cerebellum. Trend Cognit Sci 2:338–347

Chapter 3
Attention Versus Consciousness: Fused or Independent?

3.1 The Original Attack

The American neuroscientist Christof Koch, working with his group at Caltech, has spearheaded the important thesis that consciousness and attention are not as closely fused as is almost universally claimed. The usual statement is that attention to a stimulus in the external world is necessary for consciousness of that stimulus to arise in a subject. More may be needed for that consciousness to arise, but at least attention must be directed to that stimulus in order for there to be any chance of consciousness of the stimulus. This implies that consciousness is to be searched for in the interstices of attention. Koch claims that this is not the case. In particular he and his colleague Tsuchiya (Koch and Tsuchiya 2007) wrote that consciousness could arise in a subject without them paying attention to the relevant stimulus. This claim was based on several subtle experiments, which clearly we need to analyze in some detail in order to investigate this important claim.

We will do that in this chapter, where we consider in some detail how certain of these experiments may be explained, and most especially if it is the case that consciousness can arise without attention, or if instead attention is being used in a subtle manner in the process of consciousness creation.

As part of this analysis we will also consider the important process of creativity, a well-known one involving hard, conscious work to build up a good data base from which to solve a difficult problem and then release from conscious control (doing something else entirely different, for example), so as to allow unconscious creative processes to work to achieve a solution after all. This process involves attention slipping in and out of the processing, as we will discuss in more detail in the following. We consider this first and then turn to how attention and consciousness are related in certain subtle paradigms.

J.G. Taylor, *Solving the Mind-Body Problem by the CODAM Neural Model of Consciousness?*, Springer Series in Cognitive and Neural Systems 9,
DOI 10.1007/978-94-007-7645-6_3,

3.2 The Creativity Effects: Consciousness Versus Attention

We introduce the 'creativity effects', in which the known opposition of attention and creativity are applied to paradigms used in recent tests on confidence and accuracy of certain detection experiences. The manner subjects could use the creativity effects to have greater subjective experience and higher accuracy, but higher error rate during the paradigms with low attention, is used to give an attention-based explanation of several counterintuitive phenomena involving consciousness.

There have been several recent papers presenting experimental data supporting the claim that consciousness and attention are independent (Willimzig et al. 2008; Rahnev et al. 2009; Pastukhov and Braun 2007). These papers offer an interesting insight into the working of the brain under sometimes difficult conditions. In the first two of the three paradigms referenced above the paradigms are based on the use of the Posner effect of cueing attention to a place where a target may or may not appear corresponding to processing under what the authors term 'strong' or 'weak' attention. In the third of the referenced papers a set of dual task paradigms is employed to cause no attention at all to a peripheral phenomenon involving bistable perception. However a very recent paper (He and Raichle 2009) has shown, by using oscillating stimuli for the bistably perceived stimuli, that with no attention paid to stimuli at all there is no bistability. Thus we are left with the first two of the referenced paradigms to support the claimed independence of consciousness and attention.

In these paradigms the weak attention task of detecting a grid of lines on a noisy background with attention directed elsewhere (or discerning the tilt of the grid to the vertical) has been shown to lead to results arising from increased confidence above that of the strongly attended case, to increased accuracy in some trials and also to cases when there is an increased false alarm rate in the weakly as compared to the strongly attended case. These differences between the two cases and the further claims of enhanced visibility of stimuli reported by subjects in the weak attention case, have led to strong claims of attention and consciousness independence:

> The current finding adds further evidence to the emerging view that awareness and attention are supported by different neural mechanisms (Rahnev et al. 2009, p. 13)

But there must be more to this important story. How is it that consciousness, when allayed as it is in the 'weak' attention condition, appears to be more powerful in its ability to produce a sensation of visibility than in the strong attention condition? From the usual mantra that consciousness needs attention one would expect such confidence to be the other way round: the weak attention case should lead to lower visibility and perceptibility than in the strong attention case. How could this result be explained?

In the next section we present a personal experience that provided an initial explanation of these subtle effects, leading to expanding the range of brain activity being considered to the creative process and to its explanation.

3.3 A Personal Experience: The Open Window that Wasn't

Some time ago I had gone into my bedroom in the early afternoon to change my jacket, and I glanced very briefly into the adjoining en suite bathroom. I saw very clearly the open window, exactly as I expected it to be open as my wife likes to 'air the room'. However I do not like the cold air in winter making the bedroom too cold for comfort. So when I had got my new jacket out of my wardrobe and put the other one away I turned to go through into the en suite bathroom to close the window. Imagine my surprise when I found that it was not open after all! It had been closed, I then presumed, before I came into the bedroom. So why had I seen the open window, and that with great certainty and clarity? I had been 100 % certain that the bathroom window was open when I had first glanced at it.

I retired to my study with my new, warmer jacket on my back to sit and ponder what had happened. How had I been so certain of seeing the open window and that with great clarity? I had to admit that I had not stood and stared at the open window but only given it a very cursory glance when I first went into the bedroom. So had I made it all up? And on what basis would that have occurred?

I began to see that I could have taken such a brief glance of the window that I did not have time to process it properly. There was not enough time to have gained a complete view of the window, so I must have spun out the brief glimpse I had taken. But why did I seem to see the open window? And see it with such certainty? And then I remembered the 'prejudice' in my mind when I took that strange glimpse: I had thought quite clearly "My wife will have left it open – she often does – I'll go and close it after changing my jacket". In other words my view was filled in by the coloration from previous views of the open window, seen in many cases in the past and which I then had to close. It was almost part of my daily ritual to close the bathroom window. So my expectation of it being open brought the past times to mind when I had closed it, and so brought to mind a view of the open window. All an illusion, I realized, but one I had great confidence in during its occurrence. One in which my attention had been elsewhere when I had looked at the window the first time round – I was indeed thinking only of my forthcoming act to close that dratted open window!

This meditation on the vagaries of my mental processes led me to creativity, which is the example par excellence for the allaying of attention, thereby allowing the creative act to work efficiently. Attention gets in the way of creativity in this process, as it reduces the ability to find in one's memory a rapid but correct answer to a quiz question. Don't use attention, but instead just say what comes first into one's mind – that is the advice for quiz entrants. In creativity a similar reduction of attention is needed after the attentively hard work of collecting the facts of the case has been done. Attend and work hard first and then relax to let the creative juices work!

But this act of 'putting attention to one side' sounds suspiciously what the subjects in the Willimzig and the Rahnev paradigms have been subjected to in the published paradigms. In the weak attention condition they could be compared as being in the creative state. In the next section I will consider recent advances in understanding creativity that will help us to develop brain-based explanations of the effects observed in the paradigms reported by (Willimzig et al. 2008; Rahnev et al. 2009).

3.4 Creativity as Unattended Mental Simulation

Numerous attempts have been made to understand creativity. The most effective is that of (Wallas 1926), who proposed a multi-stage process being most effective for creativity. This involved at least the stages of

(a) Hard work, to build a firm data-base relevant to the problem needing to be solved;
(b) Hard attentive work to try to solve the problem directly by logical means;
(c) A period when attention is turned to something different, but yet the problem under issue is not forgotten;
(d) At this juncture several possible solutions to the problem float into consciousness, and one of these is realised as giving a solution;
(e) Finally, attention-driven work is resorted to 'close the gap' between the originally intuited but possibly incomplete solution to give a water-tight solution with all the t's crossed and i's dotted in the proof.

The implementation of these several creative processes in Wallas' model of creativity have been explored by numerous further researchers, such as (Wertheimer 1945) and in a brain-based neural model (Vandervert 2003). As outlined above, the first and second of the processes in this model involve work in preparing the ground of the problem at hand and making initial attempts to solve it. This work is under attention control with the subject conscious of gathering apposite knowledge and making initial steps to a solution. The third process is that of the incubation period. During this, attention is directed away from the original task. In the fourth process attention is suddenly switched back on in the 'eureka' moment. Finally the subject has to verify, but by hard, attention-controlled work, that the 'eureka' thought could solve the problem after all.

Thus we have three main stages of attention involvement in creativity:

1. Attention is applied and the data-base of the subject's knowledge of particular relevance to the problem is expanded maximally;
2. Attention is relaxed (directed elsewhere, for example) and the creative process is started;
3. Attention is then switched back on by the illuminating thought appearing valuable, so gaining attention, and the work of its verification is then attempted.

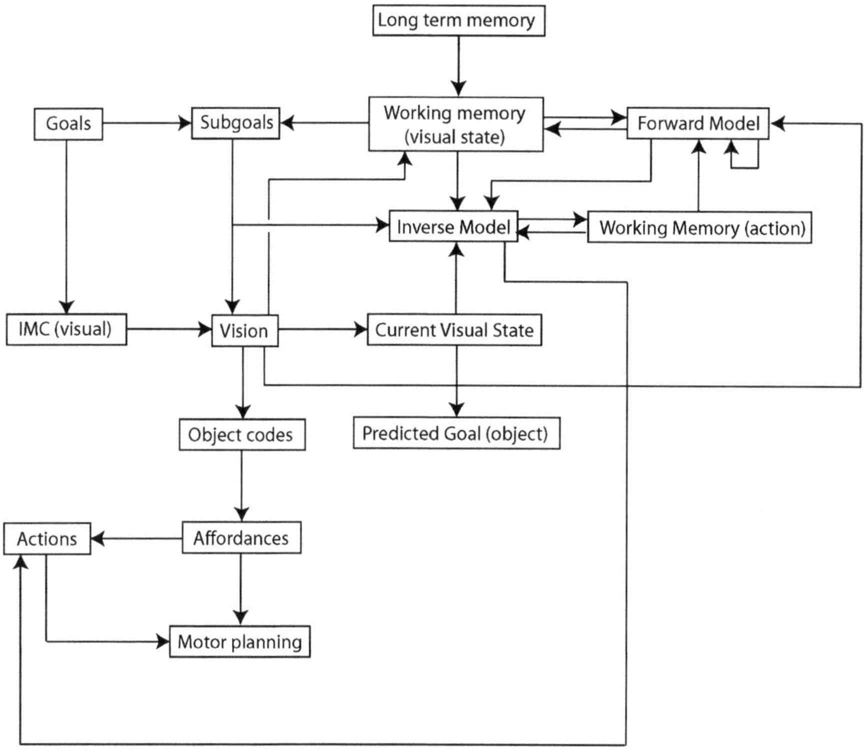

Fig. 3.1 Details of the full mental simulation loop with visual attention

We have developed a neural model of the switching of attention (Hartley and Taylor 2009). It assumes that stage 1, the work of developing the relevant database, has been completed. It allows the conscious/attended thinking process to generate a sequence of thoughts which does not reach a solution to the given problem, so is finally blocked. Attention must then be switched off, this occurring in the model by a switch triggered by the failure of the model to reach a solution (by an error assessor). The process of creativity then takes over to generate further (unconscious) thoughts, one of which finally leads to a mental state recognised as having value (say by being able roughly to extrapolate to a solution of the problem). Attention is then switched on by the reward value thereby given to this illuminating thought, so that the attended verification process can be started.

The brain basis for the architecture from which we start is that originally proposed for mental simulation (Taylor and Hartley 2008). With visual attention control added, this architecture is that shown in Fig. 3.1.

The functions of the various modules present in the figure are described in the text.

Here we briefly describe the function of each module. Details of the model neuron used are given in full in (Hartley and Taylor 2009).

Vision – This represents the basic visual input available to the model. We model this in an extremely simple manner – the region has dedicated nodes each of which holds a possible view available to the model, and they are activated depending on the simulation setup.

Object codes – Here objects already known to the system are stored, such that they can be activated by the object representation module. Each object has a corresponding node.

Goals – Here the simulation's current goal is stored and used to influence behaviour. Goals are represented by dedicated nodes. The module stores higher level end goals (such as opening a box) without specific details of how those goals are to be accomplished.

Subgoals – This region codes for lower level goals that form the components of a higher level directive – for example when opening a box subgoals might include unlatching the lid, or removing a cover. The subgoals are coded as dedicated nodes.

Affordances – This area relates objects to their uses. We suggest that object uses are coded as the affordances offered by those objects, which can then be realised through actions used on those affordances. These are coded by nodes representing these affordances (such as the affordance of opening primed by a box object).

Actions – The actions necessary to achieve affordances have representations here, which are primed by the affordances module, so the affordance of opening would activate an action representing a specific opening movement.

Inverse model – The IM is used to determine what action will achieve the goal supplied to it given the current working visual state. In a simple simulation, this output is produced by predetermined connections between a set of possible visual states and goal nodes, and output actions. These can also be learned through Hebbian style plasticity. The IM outputs the result coded as an action the system can perform

Working memory (action) – This area acts as a buffer for the actions produced by the IM, as one of a set of previously-coded action representations.

Forward model – The FM calculates the effect of a proposed action on the current state of the real or imagined world. This produces a new state. Similarly to the IM the module contains state nodes and action nodes as inputs, which are connected to state nodes as outputs, with connectivity that can either be pre-wired or learned.

Working memory (visual state) – This area holds the currently imagined visual state (as one of a set of possible states represented by dedicated nodes) produced by the forward model in a buffer such that it can be supplied to the inverse model to reach the next step towards achieving the simulated goal.

Long term memory – This allows information to be supplied to the forward model about the consequences of actions that require extra knowledge, such as imagining the room that lies behind a door and would be visible as a result of opening that door.

The attention control modules and connections in the architecture of Fig. 3.1 are as follows:

The Visual Attention Inverse Model Controller (denoted IMC (visual) in the figure): This module sends an attention feedback signal to the vision modules. Only one such module is shown in Fig. 3.1 but sensory attention can also be directed to object codes and possibly to affordances, so these modules should also be connected to the IMC (visual). The function of this attention signal is to amplify neuron activity relevant to the goal being pursued by attention (and reduce that caused by distracters).

The connection from the Goals module to the IMC (visual): This connection biases the competition assumed to be occurring on the IMC (visual) so as to allow for the attainment of attention amplification satisfying the focussing of attention on the goal object.

The connection from the Vision module to the Working Memory (visual state) module: This allows for the attention-amplified lower level activations representing the attended object to attain the visual state working memory so as to be available for report (so be in consciousness).

The connection from the Inverse model to the Action module: This connection enables the action **u** generated by the inverse model to be used to bias the Actions module representations. This causes as output from the Action module that action representation corresponding to **u**. Such a biasing process corresponds to the model of motor attention of (Taylor and Fragopanagos 2005), in which the Actions module acts as the plant in a control model in which the action IMC is the inverse model of Fig. 3.1. We see that the Taylor-Fragopanagos model is extended, in Fig. 3.1, by the presence of the forward model, thereby allowing more flexibility by possible imagined manipulations of the visual state to check for modifications a given action could produce. We emphasise that the visual state influence on the action, in both the original model and our present one, is that of the attended state in consciousness. We should add that this is not the only influence on the Actions module: there is also input from Affordances, as is expected also to influence the action being generated. This affordance may have been in consciousness (with suitable further connections to allow the affordance values to be attended to and the resultant activation attain a relevant working memory site) or not in consciousness (as in the connectivity of Fig. 3.1).

The architecture of Fig. 3.1, without the visual attention components mentioned above, was used (Taylor and Hartley 2008) to model the results of observational learning by infants in a paradigm requiring a sequence of movements to open a hierarchy of boxes contained in boxes. This was achieved in the simulation by the infant performing a mental simulation of each stage of the box openings, and then performing such action in reality. There was no need for use of the visual attention components of the architecture of Fig. 3.1 since there was only one object in the field of view and also there were only the simplest forms of internal models (the FM or forward model and IMC or inverse model controller). At no point was there need to filter out distracters, nor to allow lateral spreading of activity in object maps in order to solve difficult problems by analogy. How this latter feature of attention and its removal allow for creative thinking will be considered in the next section.

We consider the architecture of Fig. 3.1 as supporting the process of thinking at the two levels we have just described and in the introduction: at conscious and at unconscious levels. In order to switch between these levels it is necessary to consider in more detail than heretofore the visual attention components in the architecture of Fig. 3.1, especially the visual attention inverse model controller and the further attention connections included in the figure. It is through these, in concert with the other modules already present and some additional ones to be mentioned, that it will be possible to see how two levels of processing, conscious and unconscious, will be possible with the architecture.

The manner in which visual activity can be used as part of the motor control system is seen from the architecture of Fig. 3.1. The position of the Working Memory (visual state) module as sandwiched between the forward and inverse models allows there to be consciousness of the set of visual states in a mental simulation loop. That we regard as modeling the process of conscious thought (at the level of report of the content of the working memory sites).

We need now to consider how the visual states in a mental simulation loop can be taken out of consciousness but yet be part of the mental simulation loop. This can be achieved by the insertion of a switching device to allow output from the forward model to avoid the Working Memory (visual state) module, so totally avoid conscious report. This switching device will be based on an error module, as in the CODAM model of attention (Taylor 2007).

The mechanism to achieve mental simulation at a non-conscious level is by means of the connection lines in Fig. 3.1 just described, so as to avoid the Working Memory (visual state) module (and so avoid report/awareness occurring):

(a) The direct connection from the forward model to the inverse model. This enables the inverse model to produce the next action to achieve the sub-goal.
(b) The direct connection from the visual state module to the forward model. This will allow generation of the next state brought about by the new output of the inverse model and the visual state.
(c) Recurrent connection of the FM to itself if there is a sequence of virtual states to be traversed.

However more is needed to be considered in the overall creative process.

Let us turn to the example, standard in creativity testing, of giving unusual uses for an object: we take a cardboard box as an example. We can say "As a hat" as one such unusual use. That could arise from the flow of information in our brains:

Cardboard box (in picture or as words) → input processing → box nodes in object map → hat nodes in object map (by learnt lateral connections) → hat nodes in affordance map (by direct connections from the hat node in the object map and by lateral connections from the box representation of affordances to the hat representation there) → test of viability of putting on the box as a hat.

If the test of viability works, then the 'putting on hat' action becomes attended to and there is report, either by putting on the box as a hat or saying "As a hat". If the box is too large to fit stably on our head then we put it on our head and keep our hands on it to steady it; if the box is too small then we may desist from saying it could be used as a hat, or try it on as a little 'pillar box' hat.

These various responses indicate that we try out subliminally what happens if we try to put the box on our head, using the simulation loop. If successful and the action is viable then we attend to it, and hence report it. If it is not we move on to another subliminally-analyzed use.

To achieve the subliminal processing stage as well as the final report there must be an attention switch, generated as part of the IMC (visual), so that when there is an attention control signal output there is normal transmission from the forward and inverse models to their relevant working memory modules shown in Fig. 3.1. When there is no attention then the mental simulation loop circuit functions without the relevant working memory modules. It thus functions in a subliminal or unconscious manner. There will need to be an extra module for assessing the relevance of states achieved during this unconscious activation of the loop; that will be fed by the forward model in parallel with the self-recurrence (or external running of the FM) and the signal to the inverse model. Given an error-based output from this assessment module then its output would be used to bring attention to the final state and the sequence of intermediate states (assumedly not many) so as to attain the sub-goal more explicitly.

The reason for the presence of the switch itself is that of allowing the reasoning process to go 'underground' when an apparently insuperable obstacle is met by the conscious reasoning system. This may be seen as part of the extended reasoning system discussed, for example, in (Clarke 2004). However such a switching process plays a crucial role in the truly creative cognitive process. When a blockage is met in 'simpler' logical reasoning then the attention control of processing has to loose its iron grip on what is allowed to follow what in the processing, with increased reasoning and recall efficiency by subliminal-level processing. This feature is well known, for example and as noted earlier, in answering quiz questions and solving puzzles of a variety of sorts. So the switch into the subliminal mode may be achieved in the case of quizzes or creative processes such as painting or other artistic acts from the start of the search or creation process. In more general reasoning, the creative and subliminal component need only be used at points where logic gives out and more general 'extended' and creative reasoning has to step in.

In the case of our example of unusual uses of the cardboard box, the attention switch is assumed to be turned off by the goal 'unusual uses', since we know that going logically (and consciously) through a list of all possible uses of anything will not get us there, nor any other logically-based search approach. We have learnt that we need to speak 'off the top of our head', in an unattended manner. So we can regard, in a simulation of this task, that we are not using attention at all after the switch has turned it elsewhere, or reduced it to a very broad focus.

From this point of view there may well be access by the internal models during this creative phase to a considerable range of neural modules for memory of both

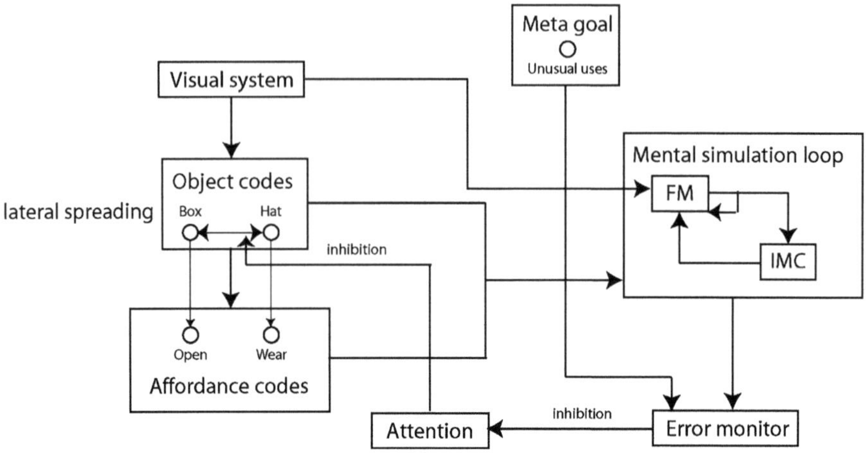

Fig. 3.2 Simplified brain-based architecture for creativity: solving the 'unusual uses task'

episodic and semantic form right across the cortex. The best approach to model this would thus be to have these connections develop as part of earlier learning processes, by such they can function initially in an attentive phase and then be useable in a subliminal one. But the presence of the unattended learning of the required lateral connections may also be possible and need to be considered.

3.4.1 Simulation Model for Unusual Uses of a Cardboard Box

To model in more detail creativity paradigms such as that involving imagining unusual uses of a cardboard box, we emphasize certain aspects of the model described above. In particular we need to allow the use of lateral spreading with object and affordance codes and look at the more specific effects of attention. We can see the architecture of the model to be used here given in more detail in Fig. 3.2 (as an extension of parts of Fig. 3.1 to handle the switch between attended and unattended processing):

The functions of the modules in Fig. 3.2 are specified in the text.

Here we detail the function of the specific modules used in Fig. 3.2:

Meta goal – The overall goal of the model is, for example, to find unusual uses of a cardboard box. As part of this process, simple action/object goal pairs are created, so we need to code both the overall goal of imagining unusual uses and the immediate goals that are tested to see if they are unusual. We have not specified the immediate goals, since we are unclear if these are used in the creative, unattended lateralisation processing. If the processing is automatic from the affordance/action codes module to the action being taken, and then used by the mental simulation loop, then no such immediate goals module is needed; that is what has been used in our architecture and simulation. If needed it can be included between the affordance codes and the mental simulation loop without any expected change in the results we report below. The meta goal is coded as a single dedicated node (with the possibility of adding more nodes for expansion, either as a distributed representation or to include other meta goals.).

Object codes – Here objects are represented by single nodes. Lateral connections between the nodes allow similar objects to be activated by this spreading, in addition to visual stimulation. It is these lateral connections which allow the use of analogy. It is assumed that these lateral connections had already been learnt during the earlier 'hard-working' attended phase of the (Wallas 1926) model mentioned in Sect. 3.1, with attention control in the lateral spreading as shown in Fig. 3.2 being learnt simultaneously.

Affordance codes – The affordance module contains nodes representing specific affordant actions that can be used on objects (such as the action of opening a box). These are primed by the object code module using pre-selected connections.

Mental simulation loop – In the full model shown in Fig. 3.1, the mental simulation loop incorporates a forward model (FM), inverse model (IM) and buffer working memories. In unattended mental simulation we suggest that these working memories are not active, such that activity passes straight between the FM and IM. The forward model generates an expected result of carrying out the action (these are pre-coded in this simplified model) while the inverse model determines the action necessary to achieve a suggested state. The function of the mental simulation loop in this simulation is to test subliminally the pairs of objects and affordances/actions generated by the lateral spreading to see if they are considered "unusual". If the use is considered unusual then attention is brought back to the system. We have not included these working memory buffers in Fig. 3.2, for simplicity.

Error monitor – The error monitor is needed to determine whether a given object/action pair tested by the mental simulation loop has fulfilled the goal criterion of being "unusual". If this criterion is met, it then activates the attention control module such that attention is restored to the goal of finding an unusual use for the box. In this simulation the error monitor compares the selected action result (passed on from the mental simulation loop) against an internally maintained list of those considered novel.

Attention – Here we use a more specific property of the attention control system than that used so far. In particular we now require the attention system to control lateral spreading in both object and affordance modules by inhibition of lateral connections. In our model, this occurs by output from the attention module stimulating the inhibitory connections present in the object code module. When attention is focused, representations will be activated singly in each region, while after the removal of attention activity can spread to similar representations (we assume that the organization of the module is such that similar objects are laterally connected). How this attentional attenuation of lateral connection takes place at the neurobiological level is indicated to some extent by studies of visual attended (in the lower order effect case). The effect thereby allows for attention to a lower level iconic memory so as to detect the noisy and otherwise hard-to-detect stimulus.

We add finally that the HO creativity effect is expected to occur when there is no stimulus present at all in the target area. On the other hand the LO creativity effect could be present when there is a very noisy target in the unattended place that can be more efficiently detected in the noise when unattended (due to the enlarged receptive fields of lower order neurons), and so arise more easily into consciousness after the stimulus has been extinguished and attention paid to the iconic memory of the stimulus.

3.6 The Creativity Effects and the Recent Paradigms on Consciousness

We now turn to analyse the results of the experiment of (Rahnev et al. 2009) in terms of the creativity effects just introduced. The paradigm uses the original Posner cueing paradigm to cause the direction to the right side of space. Any target appearing on the left side will thus be under lowered attention, termed in (Rahnev et al. 2009) 'weak' attention. There is no claim of attention being completely absent to the left side of space in this case, although the more difficult the task (such as oriented grids, for orientation detection, at a smaller and smaller angle to each other) the more reduced attention will be expected to be directed to the non-cued side in the invalid cue condition.

The main result of the experiment is that in the invalid case there are a surprisingly large number of target hits, as well as false alarms. Thus it is quoted in the paper that there were 33 % 'target present' responses in the strongly attended case, whereas under similar conditions, but now with weak attention, there were 56 % 'grating present' responses. It is necessary to ask what stimuli the responders were seeing in the 23 % of responses in the weak attention case when there was certainly no target grid. And why had the responders been so confident that there had been a target present when it had not?

A more detailed set of responses, occurring for the four different levels of grid contrast reported on in (Rahnev et al. 2009). These are (Rahnev, private communication, 2009):

$$\text{HR (cued)}: [0.37; 0.60; 0.67; 0.92]; \text{FAR (cued)}: [0.27; 0.21; 0.11; 0.14] \tag{3.1}$$

$$\text{HR (uncued)}: [0.76; 0.79; 0.91; 0.91]; \text{FAR (uncued)}: [0.74; 0.54; 0.26; 0.40] \tag{3.2}$$

(where HR = hit rate, FAR = false alarm rate).

These data can be re-expressed in terms of

Hits/Misses = 0.67, 1.5, 2, 10 (cued)
Hits/Misses = 3, 4, 10, 10 (uncued)
False +ve/True −ve = 0.3, 0.25, 0.1, 0.1 (cued)
False +ve/True −ve = 3,1, 0.3, 0.3 (uncued)

Thus for the case of the most difficult grating angle difference, in the cued case there are 2/9 as few hits/misses as for the uncued case; there are also 1/9 as many false +ve as compared to true −ve cases. This justifies the claim that there are far more correct target recognitions but also more incorrect 'guesses' in the uncued as compared to the cued case.

In particular it is the case that there is greater accuracy in its detection when a grid is present in the uncued as compared to the cued noisy case, explained by the LO creativity effect as due to the lack of attention, thereby allowing a larger field of view (and hence greater accuracy). On the other hand when there is no grid present we can suspect that there is greater creativity (by guessing, as proposed in the HO creativity effect) in the uncued rather than cued case, with the lack of attention producing less accurate results in the former than latter case. Thus both aspects of attention reduction help to explain the results of this paradigm: that of an enlarged field of view (when a grid is present but difficult to detect without attention due to background noise), as well as the pure guessing effect (when no grid is present, so only confabulation is possible). Finally we add that in the case of a grid being present in the uncued case the grid may well act as a salient stimulus, so drawing attention to itself by an attention shift.

Such explanations (as involving an enlarged field of view in the uncued noisy case with a grid present, but greater 'creative' guessing that may happen when no grid is present, again in the uncued case) are in agreement with the further results of (Rahnev et al. 2009), who show by suitable manipulation of the data that the decisions of their subjects could only arise from at least a two channel system. As they state (Rahnev et al. 2009, p. 39) "Thus we reject the hypothesis that the attended and unattended trials are managed by the same decision making system". This would be so if creativity effects are being employed.

We can expect that in the unattended case, but with a noisy grid present, with an enlarged field of view there is a better chance of detecting such a noisy target. On the other hand when there is no target present then the subject can only confabulate, using past memories as best they can (as I did for the case of the open/closed bathroom window).

We conclude that the data of (Rahnev et al. 2009) do not support the claim of the authors that they imply independence of consciousness and attention. Rather they emphasize the need to consider in what manner the paradigm needs more careful analysis.

The creativity effect explanations of the above results leads to several predictions:

1. Higher accuracy due to the LO effect will only occur amongst stimulus presentations with a very noisy un-attended stimulus.

2. Higher false alarm rates will only occur for cases when there is no stimulus in the unattended place.
3. The expectation that false positives will especially arise in a weakly attended task after a run of strongly attended cases, when detection had been especially strong. For then there will be a strong bias from the working memory containing an image of the grid being present.
4. There should be a decreasing level of false positives in a sequence of unattended stimuli, as the bias from working memory of an observed grid decays away.
5. In a sequence of alternations of presentations of unattended and attended grids there should be a relatively constant level of false positives from the relatively constant level of bias from the working memory for the attended grids in the strongly attended cases.
6. Comparing sequences of form AU, AAU, AAAU, AAAAU, AAAAAU, ... with increasing sequences of AAA ... A trials followed by a U trial, there should be increasing numbers of false positives in the U trials as there is increasing bias from the lengthening working memory as the sequence of A trials increases (although this effect may soon run out of steam as earlier and earlier A trials have increasingly smaller effect on the bias of trial responses and imagery.
7. Such a creativity effect explanation should be checked by brain imaging, especially by fMRI. In the case of 4) above there should be an increase in the activity of a relevant working memory site as the length of the AA ... A sequences increases. There should also be an important correlation (and associated causal flow of information) shown between the relevant working memory site and a decision-making site in cingulate cortex.

Similar conclusions can be reached for the data presented in (Willimzig et al. 2008). There is again a dearth of raw data, but the figures 3 & 7 in (Willimzig et al. 2008) are indicative of certain structure. These figures indicate a close similarity of the mean confidence as a function of the correct response across the two conditions of high and low attention to the stimulus. Firstly there is a natural break in both sets of figures between correct response levels of 60–80 %. For the lower correct response levels (those below 60 %) there are seen to be more points involving low attention than high attention ones (where this is taken over the different levels of difficulty employed in the paradigm). The natural break observed can be conjectured as arising from the creativity effect being used at the lower correct response levels (for both high and low attention cases). For the higher correct response points (for both low and high attention) it can be conjectured that there is enough attention in the low attention condition so as to provide about as much information as contained in the high attention condition. This explains the agreement with the results in figure 7. Those in figure 3 are slightly different in that the lower portion of the two curves (for high and low attention conditions respectively) is roughly horizontal. Again however this provides a natural break with the high correct response points, so a similar explanation can be conjectured: the lower response accuracy points of both high and low attention conditions arise from the creativity effect whilst those for the higher accuracy response values arise from increasingly noise-free attended processing.

The conclusion of this section is that there need be no relation of attention to consciousness if these two are applied to different events or stimuli. In the (Rahnev et al. 2009) paradigm in the unattended case there is highly confident consciousness of possible hallucinations in subject's minds being produced by the creativity effect in the manner specified.

8. In the uncued case with no grid present we expect to see a delayed response in time, if the above proposal is correct that the grid will act so as ultimately to capture attention.

3.7 Conclusion

The most important conclusion of this analysis is that certain psychological paradigms must be treated with care in their use in the discussion of the relation of attention, confidence and consciousness. In particular the relation between confidence and consciousness must be treated with care in paradigms such as those of (Willimzig et al. 2008) or of (Rahnev et al. 2009). If the creativity effects of either sort are being employed then there need be no relation at all between confidence and consciousness if attention to a given stimulus is low. Nor is the supposed weakness of the 'weak' attention condition relevant to the amount of attention then applied to the creatively produced illusions of the external world, although then weak attention is needed to allow the creativity effect to take off and lead to the imagined world, with high confidence, as reported in (Rahnev et al. 2009).

A final conclusion is that the creativity effects developed and applied in this chapter to various increasingly sophisticated paradigms allow for an explanation, using attention as the gateway to consciousness, of the various results reported. The interplay between consciously attended and unconscious processing in these paradigms is subtle but understandable in terms of the creativity effects introduced. The numerous claims over the past few years that 'consciousness and attention are independent' are thus seen to be false, and the original well-supported result that attention is the gateway to consciousness still stands. This conclusion is supported by further detailed analyses of a variety of other similar paradigms (also involving various types of masking) reported in (Taylor and Fragopanagos 2007).

Acknowledgements I would like to thank Prof. Lau and Dr. Rahnev for letting me see their paper (Rahnev et al. 2009) before publication as well as for helpful communications concerning other aspects of the subject.

References

Bhattacharya J et al (2009) Posterior beta and anterior gamma oscillations predict cognitive insight. J Cogn Neurosci 21(7):1269–1279

Christoff K et al (2009) Experience sampling during fMRI reveals default network and executive system contributions to mind wandering. Proc Natl Acad Sci U S A 106(21):8719–8724

Clarke A (2004) Being there. MIT Press, Cambridge, MA

Fang F, Boyaci H, Kersten D, Murray SO (2008) Attention-dependent representations of a size illusion in V1. Curr Biol 18(21):1707–1712

Friedman-Hill SR, Robertson LC, Desimone R, Ungerleider LG (2003) Posterior parietal cortex and the filtering of distracters. Proc Natl Acad Sci U S A 1999(7):4263–4268

Hartley M, Taylor JG (2009) Mental simulation, attention and creativity. Lect Notes Comput Sci 5769:40–53

He BJ, Raichle M (2009) The fMRI signal, slow cortical potential and consciousness. Trends Cogn Sci 13(7):302–309

Koch C, Tsuchiya N (2007) Attention and consciousness: two distinct brain processes. Trends Cogn Sci 11(1):16–22

Kounios J et al (2008) The origin of insight in resting state behaviour. Neuropsychologia 46(1):281–291

Pastukhov A, Braun J (2007) Perceptual reversals need no prompting by attention. J Vis 7(10):5.1–17

Rahnev A, Maniscalco B, Huang E, Bahdo L, Lau H (2009) Weakly attended stimuli produce an inflated sense of subjective visibility (University of Columbia preprint)

Taylor JG (2007) CODAM: a model of attention leading to the creation of consciousness. Scholarpedia 2(11):1598

Taylor JG, Fragopanagos N (2005) The interaction of attention and emotion. Neural Netw 18(4):353–369

Taylor JG, Fragopanagos N (2007) Resolving some confusions over attention and consciousness. Neural Netw 20(9):993–1003

Taylor JG, Hartley M (2008) Exploring cognitive machines – neural models of reasoning, illustrated through the 2-sticks paradigm. Neurocomputing 71:2411–2419

Vandervert L (2003) How working memory and cognitive modelling functions of the cerebellum contribute to discoveries in mathematics. New Ideas Psychol 21:159–175

Wallas G (1926) The art of thought. Harper, New York

Wertheimer M (1945) Productive thinking. Harper, New York

Willimzig C, Tsuchiya N, Fahle M, Einhauser W, Koch C (2008) Spatial attention increases performance but not subjective confidence in a discrimination task. J Vis 8(5):1–10

Chapter 4
Current Brain Based Models of Consciousness

4.1 Models of Consciousness

In this chapter we will review a set of models of consciousness that are presently considered the most promising by those working in the field of neurally-based models of consciousness. We specify that the models we evaluate are indeed neurally-based, since neural networks form the basic structure of the brain. However the models need not necessarily be ones expressed in mathematical formulae, ready for simulation (as we developed in Sect. 2.2 and the associated Appendix). The criterion we will apply will be in terms of the use in the models of various components of the brain, with functionality of those components being seen to be based on that of the various lobes or modules of the cortex or of sub-cortical nuclei.

The purpose of the chapter is both to review the relevant models as well as to explore which of them may or may not possess suitable characteristics to be regarded as a successful candidate, one needing more follow-up and able to support further analysis of topics such as the sense of the inner self or 'I', of its breakdown in schizophrenia, and of how consciousness itself evolved over the aeons of the last billion years or so. There are also the questions of the possibility of 'super-consciousness', such as might arise in humans by further evolution under the stresses of the weight of the human population and of the possibility of machine consciousness, although these are not discussed in any detail here but are elsewhere (Taylor 2009, 2011b).

There is presently much controversy over how consciousness is to be explained. Even amongst those who accept that it is only through brain activity that consciousness arises, there is a strong divergence of opinion as to exactly how the brain achieves this subtle and remarkable feat of 'becoming conscious'. So there are many models of how consciousness is created in brain matter. In this chapter we wish to compare and contrast the most important of these models with each other. More crucially we also present a set of natural criteria which allow us to test how well each model passes this simple set of criteria.

J.G. Taylor, *Solving the Mind-Body Problem by the CODAM Neural Model of Consciousness?*, Springer Series in Cognitive and Neural Systems 9,
DOI 10.1007/978-94-007-7645-6_4, © Springer Science+Business Media Dordrecht 2013

The criteria we consider for each model are to fit:

- All the most highly relevant data, that is data especially associated with the presence or absence of conscious experience in a controlled fashion, such as occurs in the attentional blink or other masking-type experiments (Taylor and Fragopanagos 2007)
- Data arising from the experiences of those with mental ill-health, and especially those with schizophrenia, this being a disease which most strongly affects the minds (the 'I') of the sufferers (Sass and Parnas 2003; Cermolacce et al. 2007)
- The present experimental data on attention (Corbetta et al. 2008; Gregoriou et al. 2009; Bressler et al. 2008)
- The existence and nature of the 'inner self' of Western phenomenology (Zahavi 2005)

The first and second criteria are those that would be applied by any hard-nosed scientist to any model of a part of the world – try to fit the extant and relevant data, whatever that is, either for the normal or diseased brain respectively. Both of these criteria must be applied successively and with increasing rigour so that either the model fails and a new model is to be put in its place which should fit the recalcitrant data, or it succeeds to fit the data and is available to be tested on further data. All of the models clearly have a long way to go to fit all possible and relevant data; the symbol P in the Table of the test results below indicate only a partial fit to available data. We describe in detail later what is to be considered as 'relevant' data, since it is not possible to achieve any fitting in one fell swoop.

The third criterion can be seen as also coming under the headings of tests 1 and 2. However it is of different character, involving as it does the specific mental faculty of attention. The inclusion of such a processing component is well-founded and requires insertion of a certain level of detail of the model itself. Some of the models of the list discussed in Sect. 4.2 do not possess enough detail to allow such inclusion, and so the relevant model is denoted as 'incomplete', denoted I in the table).

The fourth criterion is the most subtle of all, but yet is the most important to help 'bridge the gap' between outer and inner experience (Levine 1983), to help solve the 'hard problem' (Chalmers 1997), to give a sense of 'what it is like to be' (Nagel 1974) and also to provide the 'immunity to error through misidentification of the first person pronoun' (Shoemaker 1968). It is that component of any model that should be associated with the creation of the inner self, corresponding to the sense of 'I'. Without that experience there is no consciousness, as will be argued in many further places in this book and has been analysed most persuasively and carefully over the last century through Western phenomenology, as eloquently explained in (Zahavi 2005). We consider the criterion that such an inner self be part of the dynamics of the model supposedly crucial for consciousness creation in the brain. Only one model survives the test, that of CODAM (Taylor 2007, 2010, 2011a). However since that model constitutes the major part of the rest of this book we do not include any discussion of it here, but leave it to Chap. 6 and beyond.

4.2 Separate Models

Some of the models are not formulated so as to see how, even in principle, they would necessarily bridge the gap; however they may be able to do so in further work and so warrant consideration here.

4.2.1 The Higher Order Thought (HOT) Model

This class of models has been of great interest to those studying the philosophy of mind, and have caused (and are still causing) considerable controversy (as discussed in more detail below). It is also being employed by various working neuroscientists (see for example the chapter by Rolls 2011). Its main thesis is that a thought becomes conscious if another thought is focussed on (or is about) the first thought (Rosenthal 1986, 1993). Thus Rosenthal succinctly puts it: "The core of the theory, then, is that a mental state is a conscious state when and only when it is accompanied by a suitable 'higher order' thought" (Rosenthal 1993), The contents of that higher order thought are that the subject is in that first mental state in order for there to be consciousness of it. It is usual to consider the first thought as one at a lower level and the second, consciousness-producing thought, as at a higher (in this case second) level. Hence arises the name 'higher order theory' or HOT. Such an approach is very simple, and has attracted much discussion, as noted, some of which will be discussed below. A less restrictive variant of the above HOT model is that of (Carruthers 1996) who proposes that a state is conscious if there is a disposition to have a thought that one is in that state.

One of the main problems with the HOT approach is that it seems difficult, in order to describe consciousness by HOT in a complete manner, to avoid the need for an infinite sequence of levels of thoughts (Rowlands 2001; Zahavi 2005) or the possibility of circularity, in which one is trying to define consciousness by using it to define itself (Rowlands 2001). The former of these difficulties arises from the problem of the specification of the second order thought as knowing that one is in that lower order mental state. How can that be unless the second order thought itself is conscious (especially of being similar to the first order state). Thus consciousness is already assumed as a given in the definition of the second order thought, so leading to circularity.

Another crucial problem is that of the possibility of getting into an infinite regress by this definition. Such an infinite regress arises in HOT by the attempt to unravel the circularity just identified. Thus if the second order thought is conscious (of being in that first order mental state) then a third order thought is needed, by the HOT principle quoted from Rosenthal above, to be focussed on this second order thought to make it conscious. But then a fourth order thought will be needed to ... and so on. This leads to the infinite regress mentioned, and to the associated loss of any useful approach to consciousness. Various arguments from a dispositional viewpoint have been attempted to avoid this (and the circularity) problem but also seem to fail to be satisfactory (Rowlands 2001).

However the reason for the continued existence of the HOT approach to consciousness in spite of these difficulties is that it has somewhat of a correct 'feel' to it, and so should be present in some form in any final theory of consciousness. Such a possibility will be discussed at the end of the penultimate section.

4.2.2 The Working Memory (WM) Model

This is one of the neuro-physiologically simplest of the models of consciousness presently on offer. It is based on the idea that if neural activity in some region of the brain converges to a stable attractor level of activity, even for a limited time, then there is consciousness of such activity. Thus if the activity of a trained Hopfield network converges to an attractor A then consciousness will arise of the previous input stimulus which caused the attractor A to be set up in the network in the first place.

There is considerable support for the notion that working memory buffers form the basis for the conscious experience of content (Taylor 1999b). These modules are expanded under the Global Workspace (GW) heading in Sect. 4.2.3. Each buffer is in general expected to have its working memory character (of holding activity over several seconds in the brain) supported by some form of recurrence. This might be internal to the working memory module or involve parietal-frontal recurrent loops across a network of modules.

Such an approach of the WM model of consciousness has been pursued especially strongly in simulation by Aleksander and colleagues (Aleksander and Morton 2007) where the WM model has been extended by developing internal memories of remembered sequences of visual images and the actions of moving from one image to the next. Each of the internal visual states is coded in recurrent nets, which are able to encode stable states as well as transitions between them. Such activities are regarded by the authors as conscious.

Working memory encoding is well known to occur in the brain, and so is to be accepted as an important component of any model of consciousness creation.

4.2.3 The Global Workspace (GW) Model

A further attractive model is that of Baars as the GW model (Baars 1998, 2002). That model proposes that there exists a GW (such as a set of coupled WMs) which if suitably accessed by stimulus activity will lead to consciousness of the resulting activity. The original version of this, introduced in the late 1980s by Baars, has been modified into a more sophisticated version (Baars 2002, p. 47): "Consciousness facilitates widespread access between otherwise independent brain functions". GW theory thus supposes that there is a set of fleeting memories of which only one can be dominant at any one time, and such that the dominant one can be distributed across many areas of the brain. The GW is thus a pivotal region with global access to other brain sites; such a style of architecture is used in various computer models in terms of a 'global blackboard'. In (Baars 2002) numerous lines of experimental support were presented for the GW model.

It is to be noted that the GW model is a broad-brush approach to consciousness and does not provide what is more specifically a detailed neural architecture for the creation of conscious experience. Various attempts have been made since to develop such an architecture, as well as develop a mathematical analysis backing up the GW approach from a dynamical systems viewpoint (Shanahan 2005; Wallace 2005).

The GW approach is an appealing and basic framework that must be seriously considered as part of the overall attack on the details of the consciousness-creating brain processes. It is also to be regarded as an extension of the working memory model of consciousness to encompass all working memory sites in the brain.

4.2.4 The Complexity Models

It is interesting to note that the notion of complexity is related to the GW approach above. As Edelman and Tononi (2000) state "When we become aware of something ... it is as if suddenly many different parts of our brain were privy to some specialised subsystem .... The wide distribution of information is guaranteed." This distribution process is close to that being suggested by Baars in his GW model. However Edelman and his colleagues decided that it was more important to emphasise the 'specialised subsystem' – it was termed the 'dynamic core', involving information flow with high complexity, and so to follow the search for regions of complexity in the brain. It would be these, it could be conjectured, which would be the sites of the creation of consciousness. This is closely related to the idea of reafference or recurrent models, to be discussed in Sect. 4.2.5, in which neuronal group selection plays an important role (through lateral interactions between neuronal groups in a given module so as to select the optimally active one by means of a competition). As part of this approach there is the "key claim ... that conscious qualia are these high-dimensional discriminations." (Seth et al. 2006, p. 10799).

In this latter paper the authors considered several mathematical measures of the complexity of neural activity in a neural network: Thus neural complexity expresses the extent to which a system is both dynamically segregated and integrated, and defined more specifically as the sum of the average mutual information across all bipartitions of the system. A further measure of complexity is that of 'information integration', defined as the 'effective information' across the informational weakest link of the system, so is the minimum information bipartition. However a brain with a high value of the parameter but displaying no activity at all would still be conscious. It is difficult to calculate this measure of complexity, since the number of bipartitions increases approximately as $N \times N$. (where N is the number of neurons in the network) Furthermore there is a simple Hopfield network that has an equivalent or greater value of the parameter for any network, so implying that such a Hopfield network would be conscious.

An alternative parameter can be defined as the causal density, termed the 'cd' of the network, obtained as proportional to the number of Granger-causal interaction in the net (determined by a multi-factor time series regression analysis). A difficulty with calculating 'cd' is that it requires a dynamical analysis of the time series of activities of the neurons of the network in order to calculate, so is increasingly difficult to do this as the number of elements of the network increase (due to the increasing numbers of variables required for the Granger causality evaluation). The overall conclusion of the authors (Seth et al. 2006) is that it appears difficult to calculate any of these complexity parameters for large systems (such as the human brain).

An alternate approach, but still preserving some notion of complexity, is to use the notion of 'traffic' introduced in (Taylor et al. 2000) in a structural model analysis of brain fMRI data in a memory retrieval task. After the regions of interest (ROIs) were determined the structural model was evaluated and the number of significantly weighted lines emanating from a given ROI calculated; those ROIs with relatively large amounts of traffic were considered as possessing greatest complexity. Such an approach needs to be done for a variety of paradigms and then the average traffic per ROI calculated: those ROIs with greatest traffic would be considered putatively as involved in the creation of consciousness. Such an evaluation leaves to one side any causal notion, but a Granger causality approach could also be taken as part of the structural modeling if needed. The traffic approach avoids high complexity by reducing the network elements to ROIs, so as aggregates of neurons.

These various approaches to brain complexity are all of relevance in attempting to get a dynamical complexity handle on consciousness. However a good example of a complex module in the brain, the primary visual cortex V1, indicates already that such a complexity measure does not necessarily indicate consciousness, since experiments have shown conclusively that there is not direct consciousness of V1 activity (Leopold 2012).

4.2.5 The Recurrent Models

There is a considerable variety of such models and of neuroscientists who are adherent to them. One in particular has had considerable publicity under the terminology of the 'reafference principle' (Edelman 1992). Another is termed ART: a powerful concept of a local neuronal architecture for cortex introduced in the 1970s (Grossberg 1976). Others have also had considerable impact on the attempt to found consciousness on the presence of recurrent brain circuits (Lamme 2006; Pollen 2003).

There is an enormous amount of recurrence in the brain: any module well connected to another brain area will have reciprocal connections with it. This is particularly noticeable in the hierarchy of visual systems, where the hierarchy is seen to run both ways: up from V1 to V2 to V3, etc., as well as back down the set of visual cortices. Such connectivity has been used in developing numerous effective vision processing models. However the detailed manner as to how consciousness could thereby be created is unclear, since no associated specific neural model of consciousness, based on recurrence, has been proposed so as to generate conscious experience itself and test it.

Moreover it is correct to say that neural recurrence in connectivity between any pair of brain modules is an important part of brain processing (recurrence in connectivity seems to be the rule rather than the isolated instance between any two brain areas) so it will thereby be expected to be important to be included in the creation of consciousness as part of an overall brain architecture.

4.2.6 The Neural Field Model

We already presented a mathematically-based neural field model earlier (in detail in the Appendix to Chap. 2). Here we will treat neural field models at a high level, and assume no prior knowledge of the approach.

Neural fields come in a variety of shapes and sizes, with different levels and types of physiological basis. Thus a 'neural field' (Amari 1977; Taylor 1999b) can be defined as a continuum of neurons in a given module; this is a relatively good approximation to neural distributions in cortex provided one is not working at a level approaching the single neuron scale. It has been especially employed by (Freeman 2007) to help describe the synchronisation and phase transformations he has observed by use of surface electrodes in humans as well as lower animals. Such phase analysis in Fourier space, with its implication of interesting switches of phase, has been developed into a fascinating story by Freeman; who distinguishes between "the neural point processes dictated by the neuron doctrine versus continuously variable neural fields generated by neural masses in cortex." (Freeman 2007, p. 1021). It is through dynamical processes in the latter continuous neural fields that Freeman sees the emergence of consciousness.

An alternative definition of neural field is that of an actual field in 3-dimensional space around a set of neurons (LaBerge and Kasevich 2007). In particular the authors claim (p. 1004) "The neural basis of consciousness is theorized here to be the elevated activity of the apical dendrite within a thalamocortical circuit." This is an interesting approach to consciousness by allowing it to expand into the surroundings of an active neuron rather than have it created from purely the interactions between spiking neurons in some complex manner. It thereby allows for a continuum approach to consciousness, so fitting some aspects of our own experience.

4.2.7 The Relational Mind

This was introduced by the author in the 1970s, based on the notion that conscious experience is strongly determined by the closeness which a given percept has to past experiences in long-term memory (Taylor 1999a). The relations a given input stimulus has to present percepts held either in the earlier processing regions in cortex and/or in long-term memory was proposed, in this approach, as the basis of conscious experience. Such elicitation of past experiences and their influence on present consciousness is undoubtedly an important component of processing leading to consciousness in the brain. Yet it has to be admitted that this approach is limited only to explaining parts of experience (those biased by the past) and does not help in getting to terms with the deeper aspects of the conscious experience itself.

4.2.8 Other Models of Consciousness

There are numerous other neurally-based models of consciousness. Some of these are difficult to analyse due to lack of clarity or of being rejected by their creators. Thus the 40 Hz model of Crick and Koch (1990) has been criticized not only by other researchers but also avowed as incorrect by themselves (Gold 1999; Crick 2004).

The later approach to consciousness of (Crick and Koch 2003) presented no detailed neural model but only a general survey of overall brain processes and a set of ten selected features they regarded as important for consciousness to emerge in the brain. The most important aspect in their latest 'theory' appears to be that of requiring some neural activity 'for a feature has to cross a threshold for consciousness of that feature'. But the approach is too unclear to be usefully inserted in the WM or GW class of models, since these have additional criteria associated with attention amplification and length of time of neural activity.

Further models are considered in (Seth 2007) to which the reader is referred.

4.3 Criteria for the Review

Given such a plethora of models we have to begin to explore how well these models fit the expectations we should have of them. To do that we develop a set of criteria which are appropriate to use and which codify a range of these expectations. These criteria should apply to any model, so are beyond any intrinsic criticism of a model as developed in the brief reviews given in Sect. 4.3. The criteria are themselves chosen to emphasise what are to the author of this book the most crucial aspects of experience and are experimental data that a neural model of consciousness should satisfy. They build on standard scientific criteria built up over the last 3 centuries but are especially adapted to consciousness in terms of the paradigms that are especially selected.

In particular we consider the following criteria:

4.3.1 *Fits to Experimental Data*

Any scientific theory worth its salt must fit relevant data. It may be a difficult task to determine if such fitting is taking place, as instanced by the recent construction of the Large Hadron Collider (costing billions of dollars and involving thousands of scientists) to test the existence of the Higgs' boson, supposedly generating mass for our observable universe.

In the case of consciousness there is a large (and increasing) amount of data relevant to such testing. Indeed one can say that almost any human activity provides such data. However in general such data is too complex, so the tasks set and the measurements arising have been slimmed down to provide simpler and more useful testing domains. At the same time there has occurred the development of brain imaging machines and of single cell/multi-unit recording techniques requiring such simpler test paradigms.

Such data provides measurements of both 'outer' brain activity at different scales according to the measuring apparatus as well as 'inner' experience. Most of such inner activity involves the report of a subject as to what they saw or heard of a set of input stimuli as part of an experimental paradigm set up by the relevant experimental team. But more recently the reported experience has been of broader nature, such as of day-dreaming or similar resting processes (Kounios et al. 2009).

It is essential that any scientific model of consciousness must be able to fit or at least explain (if only qualitatively) such experimental data as noted just above. Thus data, involving bi-stable visual experiences, such as arise in the Necker cube, or processing tasks like the attentional blink (Shapiro et al. 1994), need to be given if possible a quantitative explanation. Without that it is unclear that the proposed model of consciousness is able to be acceptable as a model properly based on brain action. Let us consider some relevant experimental paradigms.

The attentional blink (Shapiro et al. 1994) was so called because stimuli were being observed by a subject at such a fast rate that when one 'target' stimulus had been detected in the fast stream of distracter stimuli then a further one – the 'second target' – proved most difficult to detect. The time lapse between the first and second such targets for greatest difficulty was observed to be about 270 ms, being the time that attention 'blinks shut', being fully occupied with the first target.

The Necker cube is composed of the edges of a cube so drawn that when it is viewed the subject will observe one corner as being towards the viewer. After a second or so there is an automatic switch so that the opposite corner of the skeletal cube now projects forward according to the viewer. Whilst this is an important paradigm, it is not dwelt on further here since it could arise from low level competitive maps between low-level visual codes for the two cube structures and so only arise in consciousness after the hard work was done. Phenomena such as the attentional blink are more important as arising from an attempted breakdown of attention control by the second target during the blink. This approach is parallel to the high-energy physics experimental methodology in breaking the particles down into their basic constituents, becoming for attention control as to how the control can be broken down, under the harshest conditions, to find out how it is put together.

Further relevant data, beyond the attentional blink, are described and simulated more carefully in (Taylor and Fragopanagos 2007). These include object substitution, in which an object, presented and then suitably masked (to be outside of awareness by the subject) is found to still be able to create a brain signal at a reasonably high level in the processing hierarchy. In particular it was observed that there is still the so-called N2 EEG-measured signal for the masked object (the N2 occurs about 180–280 ms after the presentation of the visual stimulus and is thought to represent a signal of movement of attention to the relevant hemisphere).

Another paradigm relevant to attention and consciousness is that of change-blindness. In a quantitative study of (Fernandez-Duque and Thornton 2000; see also Landman et al. 2003) stimuli of visual objects formed an equidistant set round a circle, and after a brief mask one of the objects was replaced by another. The subject had to report if this substitution had occurred or not. A cue was presented as to where the object substitution might occur either before, during or after the appearance of the mask. Data were collected on the accuracy levels of the subjects under these three different cueing conditions.

Further paradigms were also considered qualitatively, especially on blindsight (Kentridge et al. 1999), where an otherwise blind subject GY was able to employ attention cueing in his blind-field to improve his ability to detect a cue in that same field.

The bottom line for all the models is that all such data should be able to be explained by a given underlying model. That can be regarded as an extension of the Turing test for consciousness. The mechanical system purporting to be conscious is not now hidden behind a screen. The system is now being tested much more thoroughly by probing the inner workings of its ‘brain’ at the same time as listening to its report as to what it is ‘experiencing’. It must pass the test of all such results, with ‘brain’ activity being tested as well as responses being similar to those of a human. However at the present state of available models only a limited range of experimental data is being tested as a first stage (although models based on an architecture arising from a broader range of data than another such model will clearly have an advantage in explaining more general data at a further stage of model testing). The results of such testing will be considered in more detail shortly when the various appropriate tests are considered for the various models already discussed in the previous section.

4.3.2 The Presence of Attention

It is now pretty well universally accepted in neuroscience that attention is crucial for consciousness (see however the opposing view taken in Koch and Tsuchiya 2007, and of related experimentalists) which was discussed in more detail in the previous chapter, where it was found to be incorrect for all the examples so far brought forward. In any case the claimed places where attention is not needed for consciousness tend to be abnormal, so can be neglected in any attempt to consider normal brain processes and in the distortions experienced in schizophrenia. Thus various phenomena: inattentional blindness, the attentional blink (as discussed in the previous section), neglect (where a stroke causes a patient to be unable to attend to one or other side of their visual field or body and be unaware of what happens there), extinction (when a limited form of neglect takes place, with a patient unable to attend to and be aware of a stimulus, say in their left visual field if a competing stimulus is present also in their right visual field), etc. indicate that without attention to a stimulus there is no conscious experience of the stimulus itself. There may be after-effects of the input of a subliminal stimulus, i.e. one not able to gain awareness. Thus a stimulus exposed to a subject for only a few milliseconds, with no awareness of what was presented to the subject, may still lead to priming effects which can bias later processing (Bar and Biederman 1998). We have already had a detailed discussion in the previous chapter of explaining some of the unexpected phenomenon discovered in some of subtly developed paradigms. Our results were that we could explain these by the use of an attention model, contrary to the claims of most of the experimenters on the 'other side'. Further discussion of the modeling of relevant experimental data has been given in (Taylor and Fragopanagos 2007).

We thus conclude that attention acts as the gateway to consciousness. Without an attention control system as part of the model then only very simple experiences would be accessible to consciousness. In particular, only in environments with a single stimulus would a subject be able separate the sensory inputs to make any sense of them as single stimuli. Since such a criterion of single stimuli would be far too limiting for the claims made for all of the various models we considered in the previous chapter then the possession of attention would seem critical for any model. If the model does not posses this faculty at a quantitative level (not purely qualitatively) then it can only be considered as incomplete.

4.3.3 Providing an Explanation of Mental Diseases

This set of tests can be more properly considered as part of the tests under the heading of Sect. 4.3.1, since they involve experimental data both from the reports of sufferers of the various diseases and of the relevant damaged regions of their brain. In particular we consider the experiences of schizophrenics to be the example bar none for a model of consciousness to be expected to explain. For schizophrenia has the greatest set of distortions to the self amongst mental diseases. One might consider that dementia possesses such a property, although the main defects there appear more in the light of memory loss and the associated emotional disturbances this causes. However in schizophrenia the sufferer talks about 'losing my self, my I' and being 'increasingly distanced from the world' (Sass and Parnas 2003; Cermolacce et al. 2007). It is such distortions which would appear initially to be directed at the inner self as an important but vulnerable component of the sufferers' experience. Any model of consciousness should be able to attempt to explain that loss of 'I' in the schizophrenic.

4.3.4 Existence of an Inner Self

This is the most controversial of the set of criteria to be applied to models of consciousness. It relates closely to the remarks just made about the schizophrenic experience in Sect. 4.3.3, which is being increasingly and effectively interpreted as critically involving the distortion of the subject's inner self (Taylor 2010). However, aspects to be explained in any model of the creation of consciousness must especially include the experiences of the normal subject (Taylor 2011a).

The 'inner self' or pre-reflective self, as known by the Western phenomenologists (Husserl, Sartre, Merleau-Ponty and many others, surveyed in Zahavi 2005), and which we will meet many times throughout this book, is the crucial component still missing from the neural network models discussed so far for consciousness, and also from most Western analytic philosophers of Mind's discussion of the topic. The latter group regards the presence of some form of pre-reflective self or 'ipseity' as crucial. Ipseity denotes the inner self, sometimes termed the 'pre-reflective self' of Western phenomenology. In this paper we identify ipseity with the inner self, and that with the pre-reflective self, although there may be subtle differences between them according to different definitions in Western phenomenology (Zahavi 2005); these differences will be ignored here. Thus we expect a satisfactory model of consciousness to include a description of how it leads to the experience of the inner self or ipseity.

There is a useful definition of ipseity (inner self) in (Sass and Parnas 2003, p. 428): "... the experiential sense of being a vital and self-coinciding subject of experience or first person perspective on the world". Ipseity arises from a part of the self which is devoid of the components of the reflective self. This latter is composed of those characteristics of the self that can be obtained by reflecting on the characteristics of oneself, such as whether or not one has a beard or is impatient, and so forth. The pre-reflective or inner self or ipseity appears instead as content-free, and its existence provides a centre of gravity in which the ownership of ones' experiences is gathered.

Besides the claimed existence of the inner self is the proposed temporal flow of conscious experience. It is claimed to consist of three components, termed 'pretention', the 'primal moment' and 'protention' (Sokolowski 2000). The first of these, pretention, is a part of preparatory brain activity just before the experience of consciousness. The second, the primal moment, is the short period of the experience of consciousness itself. The third component of the temporal flow of consciousness, protention, involves a decaying activity record of the primal moment in the brain (as in a working memory site).

It is clear that without an inner self there is no-one to experience the consciousness of any content on a working memory site. Thus the inner self is the crucial component of the answer to the question raised by Nagel: 'what is it like to be?'. The immunity to error of (Shoemaker 1968) (I can have no error about what 'I' think) will be achieved in a situation in which the inner self has surety of experiencing exactly what it has expected. Finally the 'explanatory gap' (Levine 1983: the gap between first and third person accounts of experience) and Chalmers' hard problem (Chalmers 1997: how to get consciousness out of suitably designed matter structures) are solved by understanding in what manner the inner self interacts with the outer world through the working memory sites holding the content of experience. Thus the inner self, according to these ideas, must be a crucial component of any self-respecting model of consciousness.

In order for a model of consciousness thus to be able to solve the problems which have been plaguing it in the West (as noted above, raised by Nagel 1974; Shoemaker 1968; Levine 1983 and Chalmers 1997) it needs

(a) to have also a neural model of the inner self;
(b) to have a model of the manner in which this inner self interacts with the attended working memory containing the content of stimuli entering consciousness;
(c) to be able to define the temporal flow of activity so it agrees with the three successive components: pretention – primal moment – protention, as specified above (Sokolowski 2000).

We add that none of the models considered so far can be claimed to have an inner self in the sense used in this book. There are claims of self-monitoring in some of the models, but that is not the same as possession of such an inner self. The claim is made in (Seth 2007) that "High-level conceptual models can provide insights into the processes implemented by the neural mechanisms underlying consciousness, without necessarily specifying the mechanisms themselves. Several such models propose variations of the notion that consciousness arises from brain-based simulation of organism-environment interactions. These models illuminate in particular two fundamental aspects of phenomenology: the attribution of conscious experience to an experiencing 'self', and the first-person perspective that structures each conscious scene." However such knowledge does not provide an experiencing self at the level of the inner or pre-reflective self being discussed and developed here. The self described by Seth is that of the reflective self, corresponding to knowledge of what one looks like in the mirror and what one is doing currently. But the pre-reflective self is a monitor only of the corollary discharge of the attention control signal (as will be discussed in Chap. 6 and beyond), being a pre-amplifier of contentful input about to enter a working memory site. This is a concept foreign to the models discussed by (Seth 2007) and of those considered so far. Thus these models fail this most important of the criteria we have so far proposed: they do not include in any manner an inner self.

4.4 The Test Results

We present the test results for the application of each of the four criteria of the last section in tabular form in Table 4.1. We have, as the first column, whether or not the model/approach fits relevant data. As data we single out a set of experimental paradigms which are of relevance to the creation of consciousness: blindsight, inattentional blindness, object substitution, the attentional blink and several more (Taylor and Fragopanagos 2007). It is numerical fits to these paradigms (where numerical data is present) which are to be calculated by any of the models under discussion (as done in the paper just referred to). If the relevant model does not have a specific neural architecture to enable this to be done, and these phenomena have not even been discussed in the literature associated with the model, then the model fails on this task. In the second column are entries indicating the degree to which the model is based on attention control. The third column explores how far the model explains distortions in schizophrenic and other mental disease patients. Finally the fourth column states if the inner self can be glimpsed in any manner through the model.

We now discuss each of the entries in more detail.

Let us go through the list of the seven models whose results are presented in table I to justify and expand on the results given there.

Table 4.1 Results of the tests

Test criterion	Fits relevant data	Contains attention	Explains mental disease	Exists inner self
Model name				
1. HOT	X	X	X	X
2. WM	I	I	X	X
3. GW	I	I	X	X
4. Complexity	I	X	X	X
5. Recurrence	I	I	X	X
6. Neural field	I	I	X	X
7. Relational mind	I	I	X	X

I denotes 'incomplete' as mainly a descriptive account only occurs of the model; *X* denotes absent or NO

4.4.1 *HOT*

This model has, in the hands of (Rolls 2011 and 2007) been especially extended to the linguistic domain by means of what Rolls has termed HOST (higher order syntactic thought). He proposes specifically (Rolls 2007, p. 962) that it "feels like something to be an organism that can think about its own linguistic and semantically-based thoughts". The neural basis of this approach is possible, but still would seem to suffer from the problems of circularity and regress, criticisms already raised in our review of the HOT approach in Sect. 4.2.1. A strongly linguistic basis for HOT does not seem to avoid these earlier problems.

At the same time there is little to be seen in either the past literature on HOT or in the HOST update to put any success under the first column of Table 4.1 and similarly there is no success for the remaining columns. There is no clear explanation of distorted experiences in mental diseases. There is also a complete absence of any ability of the model to create an inner self. Altogether the model does poorly on these four tests.

4.4.2 *WM*

Reasonably clear entries arise from the review in Sect. 4.2.2: there are various models over the past of working memory applied to the phonological loop and various memory phenomena such as recency and primacy effects. Yet these paradigms are not those proposed as most relevant to consciousness (Taylor and Fragopanagos 2007), so explaining the 'I' entry. Nor have there been attempts to apply these ideas to mental diseases. Furthermore there is no hint in the WM models of how the inner self could arise from purely continued brain activity over whatever length of time.

4.4.3 *GW*

Similar features apply to the GW model as to the WM model, so explaining the entries in the table for this approach. There is a broader range of relevant experimental data which GW explains as compared to a single WM, although this is still discussed only in a qualitative manner (Baars 2002).

4.4.4 *Complexity*

Some data has been explained (on complexity) by the proponents of the approach, but no data relevant to the key experiments discussed in (Taylor and Fragopanagos

2007). Moreover there is no hint presently of attention, distorted mental disease experience or the inner self being explained through complexity.

4.4.5 Recurrence

The various approaches to recurrence mentioned in Sect. 4.2.5 have led to specific models of pattern processing in vision and improved understanding of the difficulties for such models assuming only simple feed-forward visual processing. Attention can be included if desired and many models of attention, using recurrence at various stages, inside neural models, have been developed and used in vision tasks. However this inclusion is incomplete in terms of explaining relevant phenomena such as the attentional blink and the related phenomena mentioned at the beginning of this section and associated with the entry in Table 4.1 (Taylor and Fragopanagos 2007). Mental disease and the nature of the inner self have so far been outside these model approaches emphasizing recurrence as a basis for consciousness.

4.4.6 Neural Field Theory

The neural field (as described in Chap. 2, and in some detail in its appendix) can be recognized as a framework inside which to develop a broad range of neural architectures to explain various phenomena in brain processing (visual coding, illusions of colour and other visual processing, etc.). The use of the continuum has been especially emphasized by (Freeman 2007) and important progress made thereby in understanding the more general dynamical systems approach to neural processing. However the nature of consciousness and its generation through such processing is still not completely clear, so the approach is still in the early days of development. That is why only incomplete entries are given for fitting data and attention inclusion. No progress has occurred in this model in analysis of mental diseases and the inner self.

4.4.7 Relational Mind

This approach has the same level of progress as the neural field approach (in spite of their qualitative dissimilarity), so has similar entries in the table.

4.5 Conclusions

In this chapter we have tried to compare and contrast the various main neural network models of the creation of consciousness in the human brain that are presently at the centre of considerable discussion (other than CODAM). Having briefly reviewed some of these models in Sect. 4.2, we considered how they could be assessed. In Sect. 4.3 we introduced a set of four criteria: fitting relevant data, containing attention, explaining mental diseases and finally being able to give a hint on the nature of the creation of the inner self.

On applying these four criteria we find none of the seven models considered able to be even relatively successful on all four of these counts. That does not mean that none of these models could ever be effective, but only that they presently have not necessarily reached the same stage of development. An example of this is the HOT model, which has little specific neural architecture proposed to back it up and so allow more quantitative criteria, such as in the table, to be applied. However some of the models would need considerable conceptual development to be able to reach the level of applicability to the problems of consciousness (such as those of schizophrenia) that we have considered as critical.

It may be argued that the criteria in the table are not relevant to assessing such models, especially the last one of the creation of the inner self. Even the existence of an inner self may be denied by some workers in the field of consciousness. The first three criteria of the table are part of the ongoing approach of science: to explain the relevant part of the world in a quantitative manner. It is only the last, the existence of an inner self as created by the dynamics of a given model, which does not fit that solely quantitative approach. Yet the inner self is relevant to be taken account of scientifically because it involves report. Report has been a mainstay of most paradigms in experimental psychology. Such report on the inner self is about a subtle component of experience. But it cannot be excluded from the class of inner reports, in spite of its more qualitative rather than quantitative aspect. It might ultimately be possible to bring the inner self into the quantitative universe by quantifying the level of the interaction between the inner self and the external world. If it is reduced in schizophrenia why not ask sufferers to quantify by how much their inner selves had been reduced? That may be very hard to assess but should be tried. In any case discussion of the inner self is still valid from a scientific point of view, since it would appear to have strong evidence for its existence from reports of its distortion in schizophrenia as well as from our own experience throughout each of our lives. To claim that we live without an inner self is as much as stating 'I am a zombie'. Indeed without an inner self we would indeed be in such a state. For then, as already mentioned earlier, there would be no experiencer, no-one for whom 'it is like to be'. Such is the fate of all of the models of (Seth 2007) except for the briefly mentioned CODAM model. We will come to that model in Chap. 6, with its further development in the rest of the book.

From the model viewpoint, consciousness research still has a multitude of models to contend with. Those who have proposed their own pet model strongly hold on to it, in spite of possible evidence to the contrary or internal logical difficulties. The number of citations to a given model also depends on how strongly the progenitor shouts out from the roof-tops (for that read 'conferences' and published papers) about the greatness of their model. Yet real progress will only be made by both theoretical confrontations between models and their confrontation with experimental data as presented, for example, in Table 4.1. Only through such confrontations can we hope to make real progress in the important subject of understanding consciousness.

Finally we note that CODAM, if it were included in the list of Table 4.1, would have passed all of the tests we considered above. That is a claim we will expand on in much more detail further in this book.

References

Aleksander I, Morton H (2007) Phenomenology and digital neural architectures. Neural Netw 20:932–937

Amari S-I (1977) Dynamics of pattern formation in lateral-inhibition type neural fields. Biol Cybern 27:77–87

Baars B (1998) A cognitive theory of consciousness. Cambridge University Press, Cambridge

Baars B (2002) The conscious access hypothesis: origins and recent evidence. Trends Cogn Sci 6(1):47–52

Bar M, Biederman I (1998) Subliminal visual priming. Psychol Sci 9(6):464–469

Bressler SL, Tang W, Sylvester CM, Shulman GL, Corbetta M (2008) Top-down control of human visual cortex by frontal and parietal cortex in anticipatory visual spatial attention. J Neurosci 28(40):10056–10061

Carruthers P (1996) Natural theories of consciousness. Eur J Philos 6:203–222

Cermolacce M, Naudin J, Parnas J (2007) The "minimal self" in psychopathology: re-examining the self-disorders in the schizophrenia spectrum. Conscious Cogn 16:703–714

Chalmers D (1997) The conscious mind: in search of a fundamental theory. Oxford University Press, Oxford

Corbetta M, Patel G, Shulman GL (2008) The reorienting system of the human brain: from environment to theory of mind. Neuron 58:306–324

Crick FHC (2004) The astonishing hypothesis. Cambridge University Press, Cambridge

Crick FHC, Koch C (1990) Towards a neurobiological theory of consciousness. Semin Neurosci 2:263–275

Crick FHC, Koch C (2003) A framework for consciousness. Nat Neurosci 6:119–126

Edelman GM (1992) Bright air brilliant fire. Basic Books, New York

Edelman GM, Tononi G (2000) A universe of consciousness. Basic Books, New York

Fernandez-Duque D, Thornton JM (2000) Change detection without awareness. Vis Cogn 7:323–344

Freeman WJ (2007) Indirect biological measures of consciousness from field studies of brains as dynamical systems. Neural Netw 20:1021–1031

Gold I (1999) Does 40Hz oscillation play a role in visual consciousness? Conscious Cogn 8:186–195

Gregoriou GG, Gotts SJ, Zhou H, Desimone R (2009) High-frequency, long-range coupling between prefrontal and visual cortex during attention. Science 324:1207–1210

Grossberg S (1976) Adaptive pattern classification and universal recoding, I: parallel development and coding of neural feature detectors & II: feedback, expectation, olfaction, and illusions. Biol Cybern 23:121–134, & 187–202

Kentridge RW, Heywood CA, Weiskrantz L (1999) Attention without awareness in blindsight. Proc Biol Sci 266:1805–1811

Koch C, Tsuchiya N (2007) Attention and consciousness: two distinct brain processes. Trends Cogn Sci 11(1):16–22

Kounios J et al (2009) The origin of insight in resting state behaviour. Neuropsychologia 46(1):281–291

Laberge D, Kasevich R (2007) The apical dendritic theory of consciousness. Neural Netw 20:1004–1020

Lamme VAF (2006) Towards a true neural stance on consciousness. Trends Cogn Sci 10(11):494–501

Landman R, Spekreijse H, Lamme VAF (2003) Large capacity storage of integrated objects before change blindness. Vision Res 43:149–164

Leopold DA (2012) Primary visual cortex: consciousness and blindsight. Ann Rev Neurosci 35:91–109

Levine J (1983) Materialism and qualia: the explanatory gap. Pac Philos Q 64:354–361

Nagel T (1974) What is it like to be a bat? Philos Rev 83:434–450

Pollen DA (2003) Explicit neural representations, recursive neural networks and conscious visual perception. Cereb Cortex 13(8):807–814

Rolls E (2007) A computational neuroscience approach to consciousness. Neural Netw 20(9):962–982

Rolls E (2011) Consciousness, decision making and neural computation. In: Cutsuridis V, Hussain A, Taylor JG (eds) Perception-action cycle: models, algorithms and hardware. Springer, New York, pp 287–334

Rosenthal D (1986) Two concepts of consciousness. Philos Stud 49:329–359

Rosenthal D (1993) Thinking that one thinks. In: Humphreys G, Davies M (eds) Consciousness. Blackwell, Oxford

Rowlands M (2001) Consciousness and higher-order thoughts. Mind Lang 16(3):290–310

Sass LA, Parnas J (2003) Schizophrenia, consciousness and the self. Schizophr Bull 29(3):427–444

Seth A (2007) Models of consciousness. Scholarpedia 2(1):1328

Seth A, Izhikevich E, Reeke GN, Edelman GM (2006) Theories and measures of consciousness: an extended framework. Proc Natl Acad Sci USA 103(28):10799–10804

Shanahan MP (2005) Global access, embodiment, and the conscious subject. J Conscious Stud 12(12):46–66

Shapiro KL, Raymond JE, Ansell KM (1994) Attention to visual pattern information produces the attentional blink in rapid serial visual presentation. J Exp Psychol: Hum Percept Perform 20:357–371

Shoemaker S (1968) Self reference & self-awareness. J Philos 65:555–567

Sokolowski R (2000) Introduction to phenomenology. Cambridge University Press, Cambridge

Taylor JG (1999a) The race for consciousness. MIT Press, Cambridge, MA

Taylor JG (1999b) Neural bubble dynamics in two dimensions: foundations. Biol Cybern 80:5167–5174

Taylor JG (2007) The CODAM model: through attention to consciousness. Scholarpedia 2(11):1598

Taylor JG (2009) Beyond consciousness? Int J Mach Conscious 1(1):11–21

Taylor JG (2010) The 'I' 's eye view of its consciousness. J Conscious Stud 17(1–2):95–117

Taylor JG (2011a) A neural model of the loss of self in schizophrenia. Schizophr Bull 37(6):1229–1247

Taylor JG (2011b) Can functional and phenomenal consciousness be divided? In: Proceedings of AISB 2011 on machine consciousness, University of York, 4–7 April 2011

Taylor JG, Fragopanagos N (2007) Resolving some confusions over attention and consciousness. Neural Netw 20:993–1003

Taylor JG, Horwitz B, Shah NJ, Fellenz WA, Mueller-Gaertner H-W, Krause JB (2000) Decomposing memory: functional assignments and brain traffic in paired word associate learning. Neural Netw 13(8–9):923–940

Tsuchiya N, Koch C (2008) Attention and consciousness. Scholarpedia 3(5):4173

Wallace R (2005) Consciousness: a mathematical treatment of the global neuronal workspace model. Springer, Berlin/New York

Zahavi D (2005) Subjectivity and selfhood: investigating the first-person perspective. MIT Press, Cambridge, MA

Chapter 5
The Control Nature of Attention

5.1 Introduction

We analyze in this chapter the manner in which attention acts as a controller in the brain. We hope thereby to be able to begin a more complete probe of attention and so try to uncover in what manner consciousness itself might be hidden in the interstices of some part of the attention control system.

To proceed we will consider the recently observed dorsal and ventral attention networks (DAN & VAN) and other related interacting networks, in order to clarify the manner in which attention is used in controlling decision making and response. Functionality guidance by engineering control ideas is suggested as proving important clues to help better understand attention control.

There has been important recent progress in the attempt to understand the overall architecture of the human brain, using brain imaging results as well as single cell data. In particular there is evidence that the brain is composed of several interacting networks of information processing, as analysed by functional correlational MRI (fcMRI) (Fox et al. 2005). At least six such networks have been detected by fcMRI. These results are of great importance in putting together the story of the whole brain in action, but they leave to one side the manner in which a unified response pattern of behavior and internal experience results from this overall activity. The most crucial question is whether there is an ultimate centre of control (the Central Representation described in Chap. 2, for example) in the brain or if the brain achieves its effectiveness by using a set of distributed control processes to achieve its results. However this latter devolved architecture would still need some sort of system for conflict resolution to prevent inefficiency. Thus there needs to be an ultimate conflict resolution mechanism. We consider here the hypothesis that this is based on attention and analyze its character.

Attention is known to be achieved in the brain by feedback signals to lower cortical regions from those of higher order. These signals amplify the neural activities encoding a given stimulus to be attended to and inhibit those which arise from distracters. As such there is a division of the brain into a lower controlled

J.G. Taylor, *Solving the Mind-Body Problem by the CODAM Neural Model of Consciousness?*, Springer Series in Cognitive and Neural Systems 9,
DOI 10.1007/978-94-007-7645-6_5, © Springer Science+Business Media Dordrecht 2013

region (the 'plant' in engineering control terms) and a higher-level source of the attention control signals (an inverse model controller or IMC for short, in engineering control terminology). Recent imaging and single cell results imply that such a division is supported by the flow of information in the brain (Bressler et al. 2008; Gregoriou et al. 2009), where Granger causality has been used to analyse the coupled activity in FEF, IPS and lower cortical regions (V3, for example). We will discuss these results in more detail shortly. The results imply a flow of information from the prefrontal FEF and the parietal IPS down to V3 on the development of attention control, just before the detection of a given target. Thus there is strong support for the ballistic control model of 'biased competition' of Desimone and Duncan (1995), as will also be discussed in more detail shortly.

The results of more global analysis by fMRI (Corbetta et al. 2008) have shown there are at least two networks involved in the two sorts of attention (exogenous or externally directed and endogenous, or internally directed) being investigated: the Dorsal and the Ventral networks (observed as correlated together by fcMRI). The first of these, the dorsal attention network (denoted by DAN) is the one already discussed in the previous paragraph, being composed of the FEF and IPS, and sending feedback signals from FEF, partly through IPS, to lower cortical sites (Bressler et al. 2008; Gregoriou et al. 2009). The second one is the ventral attention network (the VAN) which is involved in various aspects of attention orienting by unexpected stimuli or those in an unexpected position, usually termed exogenous attention (as compared to the top-down control of attention, termed endogenous attention).

5.2 The DAN and VAN Networks for Attention

There has already been an extensive review of the existence and relationship between the DAN & VAN fronto-parietal attention control networks and the extent they are employed in various attention paradigms (Corbetta et al. 2008); we do not intend to duplicate that in this chapter but refer readers to it for completeness. In summary, the DAN network (composed of FEF and IPS/SPL cortical regions) has been shown to be involved in directing attention to spatial processing, and the VAN (composed of TPJ and IFG cortical regions) is an additional component activated in addition to DAN in the case of unexpected but target-relevant stimuli appearing in an unattended place in the visual field. It has also been suggested that VAN activity may break through into the DAN-controlled endogenous attention network if an important enough stimulus enters the brain, as in exogenous attention. The mechanism by which such target-based processing is carried out by the DAN has already been hinted at in the previous section, following the experimental results of (Bressler et al. 2008) and (Gregoriou et al. 2009): by means of signals generated in the higher order sites of FEF and IPS/SPL and fed back to lower-level visual (and other modality) cortices to amplify appropriate stimulus-based neural activity there. We note again that both the DAN and VAN are separately well-connected networks as observed through fcMRI (Corbetta et al. 2008).

From the references quoted in (Corbetta et al. 2008) endogenous attention orienting (defined as being guided by top-down goals, supposedly held in PFC) is achieved only by activity in the DAN. The observed components of FEF and IPS/SPL can correspond respectively to the goal bias and the inverse model controller (IMC or generator of the signal for the movement of the focus of attention) in the ballistic-type 'biased competition' model of (Desimone and Duncan 1995). This model corresponds in engineering control terms to the simplest form of control, with no feedback control or use of a predictor. Such a simple model of attention was updated in the CODAM model of (Taylor 2000, 2007) by the presence of a corollary discharge feeding a forward or predictor model, as we will discuss in more detail in the next chapter. There are also other components suggested as present, such as a buffer working memory module and an error computing module, all being present for greater efficiency of attention control. We have to consider the manner in which exogenous attention control can be fitted at a similar level of simplicity to the DAN-ballistic model before we can expect to extend any more general control model, such as the CODAM-type of more general control model, to the whole of attention control (which we do in the next chapter).

For the case of the sudden appearance of a target stimulus in an incorrect position, as in the invalid case of the Posner paradigm, there is activation in both the DAN (FEF & IPS/SPL) and the VAN (IFG and Anterior Insula), as shown, for example by (Indovina and Macaluso 2007); this activation occurs even if the exogenous target stimulus is not very distinctive. However the authors also showed that an unattended stimulus which was not task-relevant did not activate the VAN. Nor does exogenous orienting to a cue activate the VAN (even though attention may be drawn to that position by the DAN, as shown by its activation as well as speeded processing at the relevant site) (Kincade et al. 2005); indeed these authors found that the DAN activation in the case of a salient exogenous cue was greater than for a neutral cue, with strengths of the DAN activation for the three types of cue having an ordering of responses as:

$$\text{neutral} < \text{exogenous} < \text{endogenous}$$

However the VAN is activated by irrelevant objects when they posses some similarity to a target object (Serences et al. 2005). As summarized in (Corbetta et al. 2008, p. 310) "...the ventral network is not activated by orienting to distinctive but unimportant stimuli (exogenous orienting) ... but does underlie reorienting to environmental stimuli based on their task relevance."

Thus the DAN orients to both exogenous and goal-based stimuli, whilst the DAN & VAN are both involved in stimulus-driven reorienting.

It is still necessary to ask how we can fuse the DAN and VAN in an overall ballistic-style model of attention orienting. In order to develop such a model let us consider the relation between the DAN-based IMC module of IPS/SPL and a possible additional IMC in the TPJ module. We are led to the possibility that both IPS/SPL and TPJ could function as distinct IMCs directing their own part of the overall attention control circuitry. As part of this we must take especial notice of the presence of MFG which, as noted in (Corbetta et al. 2008), is the crucial connector between the DAN and VAN.

If we assume that attention's main role is to amplify lower level cortical activations for an attended stimulus by top-down feedback, it does seem possible to allow for two (or more) IMC-like control modules. Using the searchlight analogy of attention it would be as if there were two searchlights probing lower visual cortices: one might think that there is no reason why that should not arise (as occurs in military operations, especially with several searchlights attempting to probe for an incoming intruder). However in the invalid case of the Posner paradigm, attention must be re-directed to a new position. Thus the attention signal from the DAN control system (specifically from the IMC composed of the modules IPS and SPL) requires over-riding to redirect attention to the position where the new target arises. In other words it is necessary to have an overall controller to redirect attention to the correct position. In the searchlight analogy the two possible searchlights introduced above must somehow be coupled together so as to achieve a correct re-direction of attention to the same actual target, whenever it is the most salient. In other words both searchlight beams must be directed to the same position, instead of pointing at two completely different places (for example one searchlight to the invalid site, the other to the new correct site in the invalid case of the Posner paradigm). This is the problem of breakthrough into attention needed by an exogenous target or new but attention- demanding object: how does the previous attention focus (under the attention control of the IPS/SPL IMC feedback signal in the DAN) get redirected specifically by the extra activity in TPJ arising from the new target-relevant stimulus? According to figure 5 of (Corbetta et al. 2008) there is no direct connection between TPJ and the IPS/SPL complex. Any such connectivity would seem to go through MFG, described in (Corbetta et al. 2008) as the connecting link between the DAN and VAN.

One difficulty with depending on this connecting link is that noted in (Corbetta et al. 2008): MFG is de-activated, along with TPJ, when distracters only were visual stimuli in two experiments (Shulman et al. 2002; Todd and Marois 2004). In the first of these references, it was reported that subjects were presented with a rapid serial visual presentation of sets of stimuli, each for only 45 ms, composed either of only letter distracters or occasionally including a number target which had to be reported. The second paper reported on an experiment that required subjects to remember a visual display and then match it to a later probe display. TPJ showed a de-activation proportional to the number of display items that had to be retained in the second case. In the first experiment, however, the dorsal ACC and the anterior insula (AI) were both activated. These results imply that the VAN (and so the MFG) does not play a role in such paradigms, and that further control components are needed (as proposed in (Dosenbach et al. 2008, or in the CODAM extension of the DAN, Taylor 2000, 2007)).

To proceed, we notice that there are at least three sets of attention control conditions and their related observed module activations which need to be discussed through the activations of the DAN and VAN architecture:

(a) Only distracters are present, when it is observed that the modules in the VAN are inhibited, but those in the DAN are active in the case of a salient distracter (related to the reorienting of attention to the salient distracter when no target is present, for example);
(b) Only a target is present in the expected position, when the VAN is inactive and the DAN is active;
(c) An unattended target appears as a visual stimulus (as the invalid case of the Posner paradigm), when both the VAN and DAN networks are active to re-orient attention to it, and further components such as the AI and ACC are also observed active.

We will use the various observed active and inhibited modules as described so far (taken from Corbetta et al. 2008) to create a simple architecture that has activations agreeing with those observed.

In order for case (a) to occur, an inhibitory signal must be created by some form of monitor that assesses the distracters in relation to the present target. We can take the relevant monitor to be the anterior cingulate cortex (ACC), since this is both a well-known error-correcting module, as being observed active during such processing (Corbetta et al. 2008). Thus this inclusion brings in the highest order control circuit of (Dosenbach et al. 2008). For case (b) we can take the original DAN biased competition architecture of (Desimone and Duncan 1995), with no activation of the VAN. Case (c) is more complex: we do not have any clear evidence as to how the unattended target object ultimately grabs attention to be directed to it. This particularly involves the shifting of the attention focus from its initial invalid position to that which is valid. However we can follow (Corbetta et al. 2008) and (Indovina and Macaluso 2007, especially in their figure 2) to suppose that the VAN network is a circuit breaker for the DAN activity, so as to cause the attention focus signal from the IPS/SPL DAN IMC to be shifted to the correct position rather than being focussed on the invalid place. This would be achieved through a strong enough signal from the VAN network (especially TPJ) to the DAN IMC.

Consider the invalid case of the Posner paradigm. Attention has been cued to the invalid side of space. The target is suddenly shone in the opposite hemifield: how does the subject move their attention to the invalid position now occupied by the target? We assume the new visual input is processed in an unattended manner by the hierarchy of visual feature analysers (V1, V2, V3, V4, etc.). At some point in the brain's processing of this input it must acquire enough salience to draw attention to itself, overcoming that drawn by the hemisphere of the invalid cue direction. This could occur by the target stimulus reaching the ventral frontal components of the VAN, acting as an alternative goal site (possibly more object-based than FEF and less spatially-coded, as in IFG in VAN, observed in Indovina and Macaluso, their Table 1 and fig 2).

By some still-controversial mechanism (as discussed in Corbetta et al. 2008) such activation in the VAN (and associated with possible concomitant activity in the DAN) would lead to attention movement to the opposite hemisphere and hence to the previously unattended target. At that juncture the subject would become aware of the events occurring in the DAN and VAN and thereby respond in whatever way required of the paradigm. Such activation agrees with that reported in (c) above and in (Indovina and Macaluso 2007). One way to achieve such VAN-DAN cortical interaction (guided, we also assume, by sub-cortical mechanisms) is by the anterior goal components of the VAN (in IFG) sending a strong signal to TPJ and thence a large enough breakthrough signal to the DAN; such signal transmission would be by the MFG (as proposed in Corbetta et al. 2008, and postulated above), although there might be a more direct route not observable by fcMRI. However the results of (Indovina and Macaluso 2007) for the invalid Posner paradigm case do not indicate any MFG activity, but do that for INS and pre-SMA. We assume that latter areas are playing a similar role to ACC, as error monitors. Thus a route other than through MFG must be used for activity to travel from TPJ to IPS/SPL to provide the necessary attention bias to achieve breakthrough.

5.3 The Simplest Attention Control Architecture

A simple possible architecture for further more detailed exploration is proposed as that of Fig. 5.1, which is based on figure 2 of (Corbetta et al. 2008). As can be seen it is composed of a pair of ballistic control circuits coupled through MFG and the ACC/SMA modules.

The MFG acts as a connecting module between the DAN and the VAN (Corbetta et al. 2008). Such a connection route is supposedly inhibited by the error signal from

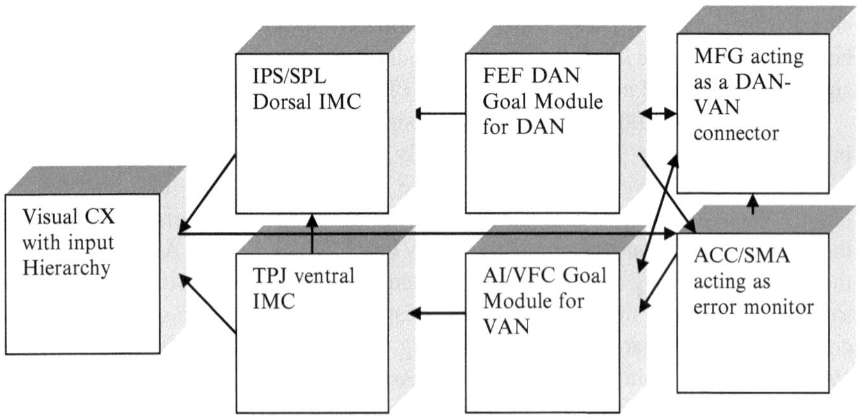

Fig. 5.1 Initial DAN + VAN attention control circuitry

ACC/SMA. The visual cortical input (entering the architecture through the extreme left module) represents both the input region of retina and thalamic geniculate nucleus as well as the hierarchy of visual cortices V1, V2, V3 … etc. The visual input (suitably processed) is sent to the ACC/SMA, acting as an error monitor, to compare this input with that coming from the goal module in FEF. If there is an error then an inhibitory signal, assumed to be generated by this error, is sent from the ACC/SMA to the AI/VFC goal module for the VAN, to the MFG as well as to the TPJ as putative IMC for the VAN. The error could be due to the presence of distracters only in the visual field, as in case (a) above. For case (b) the goal (from FEF) and input signals are identical so there is no output from ACC/SMA, so that the VAN will not be activated. Finally for case (c) activation arises in the VAN as well as the DAN, and these thereby achieve reorienting of the focus of attention to the target (such as by the VAN activity of TPJ over-riding the focus of attention in the DAN). The arrow heads give the main direction of flow of activity, although there will also be information flowing in the opposite direction, for example so as to alert the FEF goal in the VAN in case c). Moreover there is no distinction in the DAN of left and right hemispheres, which are fused together in the figure.

We note that already there are numerous predictions from the architecture of Fig. 5.1 which need exploration. The circuit shown indicates many causal flows of information from one module to another that could be checked along the lines of (Bressler et al. 2008; Gregoriou et al. 2009). The simplest approach to these predictions is to use fMRI and Granger causality (which works by constructing time delayed linear regressions models from one module or voxel to another, and the reverse, with the causality direction determined by the best fitting regression model), as in (Bressler et al. 2008) on the various paradigms discussed in this chapter so far, and especially the invalid Posner paradigm. Results of such an approach, using Granger causality to analyse the flow of information in attention, are shown in Fig. 5.2 (Bressler et al. 2008).

As stated by (Bressler et al. 2008, p. 10058) "The frontal eye fields (FEF) modulated the intraparietal sulcus (IPS) and both regions modulated the visual occipital cortex." This latter is a crucial component implying the correctness of the flow in the DAN network. It corresponds to a ballistic control system, with bias arising from the FEF to guide the direction of the feedback attention signal from the IPS down to the visual cortex.

In the case of the DAN/VAN overall network, such a causality analysis would clarify the information flow from the VAN to the DAN and answer the question: is there a flow from IFG to TPJ, then up to ACC/SMI and thence to IPS/SPL, as predicted by the architecture of Fig. 5.1 and the discussion above? Such a linear flow is expected, however, to be too simplified and a more general interaction will very likely occur, in which the error monitor function proposed for ACC/SMI will require considerably more interaction with IPS/SPL to achieve redirection of the feedback attention signal to lower level cortex to reorient attention to the position of the invalid target.

We summarise the various aspects of attention and the related components involved in Table 5.1.

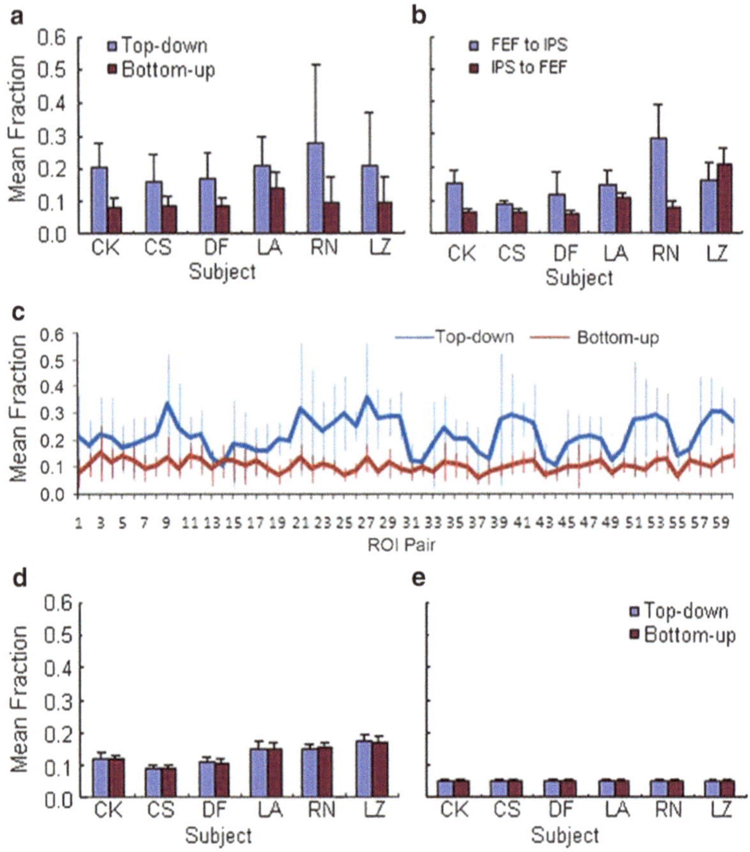

Fig. 5.2 (**a**–**e**) *Top-down* versus *bottom-up* Granger causality. The mean fraction of significant *F* (a statistical measure of significance) is significantly greater for *top-down* (*blue*) than for *bottom-up* (*red*) Granger causality when measured separately for each subject (**a**) and for 59 of 60 Region of Interest (ROI) pairs with all subjects combined (**c**), but not in the voxel-randomized (**d**) or trial-randomized (**e**) data. *Error bars* indicate variability across ROI pairs (**a**, **b**, **d**, **e**) or across subjects (**c**). **b**, The mean fraction of significant *F* is significantly greater from FEF to IPS than from IPS to FEF for five of six subjects. **c**, ROI pairs (Adapted from Bressler et al. 2008, with permission)

In the table only four attention paradigms are considered, these being the main ones for which brain activations have been reviewed in (Corbetta et al. 2008). It is these which have been used in columns 2 and 3 of the table. The entries in columns 4 and 5 summarise those proposed earlier in this section, in association with the architecture of figure 1.

The architecture of Fig. 5.1, supported somewhat by the results of (Bressler et al. 2008) presented in Fig. 5.2, still only uses ballistic control throughout, however. Why should it be extended? An important reason is to improve its efficiency: evolutionary pressure would lead us to expect that if there were a more efficient

Table 5.1 Observed excited and inhibited modules in various attention paradigms together with proposed IMC and Goals modules for the relevant attention control

Paradigm	Active modules	Inhibited modules	Proposed IMC	Proposed goals
1. Posner valid	FEF, IPS/SPL (DAN) + ACC/SMA	TPJ, VFC/AI (VAN)	IPS/SPL	FEF
2. Posner invalid (re-orientation) (Indovina and Macaluso 2007)	DAN + TPJ, VFC/AI (VAN)	MFG?	TPJ –> IPS/SPL (either directly or indirectly through MFG)	FEF + VFC/AI
3. Distracters only (Serences et al. 2005)	DAN + VAN	None detected	TPJ –> IPS/SPL (again directly or indirectly)	FEF + VFC/AI
4. Unexpected salient stimulus (Kincade et al. 2005; De Fockert et al. 2004)	DAN only (captures attn)		IPS/SPL	FEF

attention control architecture then such would have arisen through evolutionary pressure. An animal with the most efficient attention system would be expected to survive best in the competitive society of evolving animals on the earth over the last 3 billion years, since life began. The second reason would be to be able to fit present data on attention most effectively. It has been suggested over the last decade that such an extension of attention control circuitry is necessary (Taylor 2000, 2009; Fragopanagos et al. 2005; Fragopanagos and Taylor 2007).

5.4 The Possible Sites of Short-Term Memory

An important component in any extension of the control model of attention of Fig. 5.1 is that associated with the presence of a working memory system able to hold attended activity (suitably coded) for an extended time so as to make this activity accessible to other processors in other parts of the brain (such as in different modalities or in related attended stimuli). To achieve suitable understanding we need to consider the possible modules for working memory on their own, especially in terms of experimental evidence. We first consider the question as to what really is a buffer working memory and then the experimental evidence as to where in the brain such modules are sited.

It is noted in (Xu and Chun 2006; Todd and Marois 2004) that there is strong evidence of the presence of short-term memory (STM) sites in parietal lobe, from the presence of extended activations during a working memory task which are load-dependent up to a certain level, that of the capacity of the buffer. In (Yoon et al. 2006) it is shown that rehearsal of material for a later test activates PFC and VAC, with the PFC causing rehearsal commands and VAC containing details of the material to be rehearsed. A further recent study by (Kawasaki et al. 2008) showed that the activity in posterior SPL was memory-load dependent for up to three feature types (colour, shape and motion), and it was conjectured that the region collected and stored information from the dorsal and ventral visual processing streams. An even more recent study in (Mantini et al. 2009) has shown in addition that the PPC has regions holding activity in a manner increasing with load beyond the usual 3–4 items capacity limit, so such regions would be involved in rehearsal as well as passive storage. Thus the initial storage areas were suggested as being elsewhere than in PPC.

Those areas were described in (Awh and Jonides 2001; Curtis and D'Esposito 2003; Lepsien and Nobre 2005, 2007). In particular, in (Lepsien and Nobre 2005) the architecture of the brain involved in retro-cueing to cause orienting to spatial stimuli was observed by suitably effective paradigms. In these the initial stimulus array was presented and assumedly entered the subject's STM. A retro-cue was then presented a little later (2–12 s) to direct attention to an appropriate position in the earlier array (or not, in the case of a neutral cue). Finally a probe (one of the original coloured crosses or a completely different one) was presented that had to be responded to as present or absent by a suitable button press. Orienting of attention to the stored STM activity from the initial stimulus array was observed, using a 3 T fMRI scanner, to be achieved by a network that is the same as that used in orienting to external cued stimuli: bilateral PPC (IPS/SPL/PreC), Insula and Lateral and Medial PFC. In the case of retro-cueing to cause orienting attention to object stimuli (faces and scenes) it was shown in (Lepsien and Nobre 2007) that the attention reorienting was achieved by the left hemisphere DAN mentioned earlier (IPS, SPL/precuneus and DLPFC), cingulate sulcus and posterior IFS, the latter also in the left hemisphere. The continued activity was held in STM areas in the Fusiform Gyrus (for faces) and in the Parahippocampal area (for scenes).

The fact that PFC areas do not function as sole sites of STM has been strongly argued by (Postle 2006). In that paper he amassed references to many experimental results indicating the absence of STM powers in patients with PFC deficits of a range of sorts, as well as the lack of input specificity in observed PFC delay activity. Postle argued that the standard model (with PFC functioning as the site of STM) needs be modified so as to involve posterior cortical sites as those of the actual site of STM powers (coded for the relevant stimuli), as is the FFG for faces and the parahippocampal place area PPC for scenes. These earlier STM sites are under attention control, so that models such as the CODAM model is consistent with the revised approach to working memory of (Postle 2006), and the data he presents to support his arguments.

There is a further question over the manner in which there is access to these buffer working memory sites and hence to help in the search for where these are based. An important question to be raised about the buffer is as to it being only accessed by sufficiently attention-amplified stimulus activity, or is there always low-level access to the buffer but only suitably strong activity of the attended stimulus activity is able to be employed in higher-level processing? This is related to the fate of any unattended stimulus: does it reach its buffer site? This is equivalent to asking if there is early or late selection of the stimulus activity occurring (Yi et al. 2004)? In the late case then unattended activity will be assumed to have some record in the buffer, although not as strong as for the attended stimulus. The results of (Yi et al. 2004) indicate that as the attended task becomes harder (a face detection task increases in hardness due to noise on the face picture) then the unattended activity in its buffer will decrease. This is consistent with less attention being available for processing any background scene. As pointed out in (Yi et al. 2004) their results indicate that attention is not a unitary concept involving different components but that the type of attention engaged "determines the extent to which unattended visual events leave stimulus-specific neural traces in high-level perceptual cortex" (Yi et al. 2004, p. 995).

Given such results and other similar ones on the fate of unattended stimuli, we conclude that there will be numerous lower levels of activity in any input buffer associated with codes for unattended stimuli, but only the attended stimulus will have a relevant code activated at a suitably high level to be able to access other buffer and relevant prefrontal sites. Given this proviso we can safely assume that the input buffers for space, faces and scenes are the PPC (Posterior parietal cortex) and higher order visual cortices, FFG (face fusiform gyrus) and PPA (parahippocampal place area) respectively. This fits with the assumed initial extension of the CODAM model from the original ballistic attention model of (Desimone and Duncan 1995) to include working memory buffer sites.

5.5 Conclusion

We have presented strong evidence that attention is a control system. There is much use of control system ideas in engineering and many applications of such ideas across industry. However many of these applications are far more sophisticated than the model of attention we have arrived at above. In particular they employ various additions to such a simple control model, both to speed up more efficient control as well as providing a method for error correction during the generation or use of the underlying control signal. Both of these aspects are of great importance in many control applications. Hence we consider them next so as to see how employment of such ideas could also be used to make attention as more efficient. Any such improvement is expected to be used to gain advantage in the competitive battle for survival in the real world. We thereby arrive at CODAM, the central theme of this book.

References

Awh E, Jonides J (2001) Overlapping mechanisms of attention and working memory. Trends Cogn Sci 5(3):119–126

Bressler S, Tang W, Sylvester CM, Shulman GL, Corbetta M (2008) Top-down control of human visual cortex by frontal and parietal cortex in anticipatory visual spatial attention. J Neurosci 28(40):10056–10061

Corbetta M, Patel G, Shulman GL (2008) The reorienting system of the human brain: from environment to theory of mind. Neuron 58:306–324

Curtis CE, D'Esposito M (2003) Persistent activity in the prefrontal cortex during working memory. Trends Cogn Sci 7(3):415–423

De Fockert J, Rees G, Frith C, Lavie N (2004) Neural correlated of attentional capture in visual search. J Cogn Neurosci 16(5):751–759

Desimone R, Duncan J (1995) Neural mechanisms of selective visual attention. Ann Rev Neurosci 18:193–222

Dosenbach NUF, Fair DA, Cohen AL, Schlaggar BL, Petesen SE (2008) A dual-networks architecture of top-down control. Trends Cogn Sci 12(3):99–105

Fox MD, Snyder AZ, Vincent JL, Corbetta M, Van Essen DC, Raichle ME (2005) The human brain is intrinsically organized into dynamic, anticorrelated functional networks. Proc Natl Acad Sci U S A 102:9673–9678

Fragopanagos N, Taylor JG (2007) Private communication

Fragopanagos N, Kockelkoren S, Taylor JG (2005) A neurodynamic model of the attentional blink. Cogn Brain Res 24:568–586

Gregoriou GG, Gotts SJ, Zhou H, Desimone R (2009) High-frequency, long-range coupling between prefrontal and visual cortex during attention. Science 324:1207–1210

Indovina I, Macaluso E (2007) Dissociation of stimulus relevance and saliency factors during shifts of visuospatial attention. Cereb Cortex 17:1701–1711

Kawasaki M, Watanabe M, Okuda J, Sakagami M, Aihara K (2008) Human posterior parietal cortex maintains color, shape and motion in visual short-term memory. Brain Res 1213:91–97

Kincade JM, Abrams RA, Astafiev SV, Shulman GL, Corbetta M (2005) An event-related functional magnetic resonance imaging study of voluntary & stimulus-driven orienting of attention. J Neurosci 25(18):4593–4604

Lepsien J, Nobre AC (2005) Directing spatial attention in mental representations: interactions between attentional orienting and working-memory load. Neuroimage 26:733–743

Lepsien J, Nobre AC (2007) Attention modulation of object representations in working memory. Cereb Cortex 17:2072–2083

Mantini D, Corbetta M, Perrucci MG, Romani GL, Del Gratta C (2009) Large-scale brain networks account for sustained and transient activity during target detection. Neuroimage 44:265–274

Postle BR (2006) Working memory as an emergent property of the mind and brain. Neuroscience 139:23–38

Serences JT, Shomstein S, Leber AB, Golay X, Egeth HE, Yantis S (2005) Co-ordination of voluntary and stimulus-driven control in human cortex. Psychol Sci 16(2):114–122

Shulman GL, d'Avossa G, Tansy AP, Corbetta M (2002) Two attentional processes in the parietal lobe. Cereb Cortex 12(11):1124–1131

Taylor JG (2000) Attentional movement: the control basis for consciousness. Soc Neurosci Abstr 26, 2231#839.3

Taylor JG (2007) CODAM: a model of attention leading to the creation of consciousness. Scholarpedia 2(11):1598

Taylor JG (2009) Modelling consciousness. New Math Nat Comput 5(1):31–46

Todd JJ, Marois R (2004) Capacity limit of visual short-term memory in human posterior parietal cortex. Nature 428:751–754

Xu Y, Chun M-M (2006) Dissociable neural mechanisms supporting visual short-term memory. Nature 440:91–95

Yi D-J, Woodman GF, Widders D, Marois R, Chun MM (2004) Neural fate of ignored stimuli: dissociable effects of perceptual and working memory load. Nat Neurosci 7:992–996

Yoon JH, Curtis CE, D'Esposito M (2006) Differential effects of distraction during working memory on delay-period activity in the prefrontal cortex and the visual association cortex. Neuroimage 29(4):1117–1126

Chapter 6
The Full CODAM Model

6.1 Extending Attention Control by CODAM

The CODAM network was designed in the late 1990s (Taylor 1999, 2000) so as to bring attention into the modern engineering control domain by updating it beyond ballistic control by the introduction of further components, such as by the presence of a predictor or forward model and of an error corrector. It also possesses a working memory module to hold the amplified input representing the attended stimulus so as to allow it to be available for general report around the brain for further processing. Such a process of singling out a stimulus for further work is the essence of attention acting as a filter. In a complex environment, with many distracters, such a filtering action is crucial to simplify the 'world in the head' so allowing it to be used in an efficient manner at a higher level. Thus attention can be seen as a way to achieve simplified conscious report, without having to go through a possibly long list of external but unimportant stimuli for any task at hand.

We have already introduced an error monitor in Fig. 5.1, identified with ACC, and indicated how it might be employed in some attention paradigms. No forward model has been included in Fig. 5.1 however, nor is it clear how it would fit into the architecture there. Yet it was proposed (Taylor 1999, 2000) that such a module – acting as a forward or predictor model – be introduced as possibly sited in parietal lobe, and so allowing an early error signal to be created by using the forward model. Comparison of the predicted result with a required goal (supposedly held in prefrontal cortex) would then lead to any error detection. This would allow for fast error correction and so improve the efficiency of attention control. Such a fast control mechanism was employed effectively in the explanation of the attentional blink in the simulation reported in (Fragopanagos et al. 2005).

A third reason (besides the two already mentioned of increased efficiency of processing and fitting extant data) is that attention control has some characteristics of motor control, as the pre-motor theory of attention suggested by (Rizolatti and Craighero 2010). We consider these three reasons next.

J.G. Taylor, *Solving the Mind-Body Problem by the CODAM Neural Model of Consciousness?*, Springer Series in Cognitive and Neural Systems 9,
DOI 10.1007/978-94-007-7645-6_6, © Springer Science+Business Media Dordrecht 2013

There are regions in parietal cortex which are known to be crucial in attention control, since if they are de-activated by TMS, say at 200 ms or so after a stimulus onset, then a reduction of attention efficiency results. One of these regions is SMG, another is ANG. As reported in (Chambers et al. 2004a, b) both of these regions are important in attention processing. Thus there are parietal sites which are involved in early information processing (at around 200 ms and at even earlier times) for activity associated with the movement of attention focus. To probe these and other possibilities further we begin a step-by-step development of attention understanding over the next few pages. By this we will consider how we might make the attention system increasingly efficient, as might be expected if it was created through an evolutionary path over past aeons.

6.1.1 Steps 1 and 2: Modelling Attention

Presently a consensus of neuroscientists agree that consciousness is attention-based. We explored that in Chap. 3 and showed that claims to the contrary were on very shaky ground. Thus we need to start by exploring attention as thoroughly as possible and hope that in the process we will discover neural activity that can be seen, from its character, as supporting conscious experience. In particular there needs to be activity supporting the experience of the inner self. There are numerous models of consciousness which have been proposed (see for example papers cited in Taylor et al. 2007), but so far only CODAM (Taylor 2000, 2007) has a basis in attention and is seen to have room for the inner self.

In order to give a simple introduction to CODAM, we propose to analyze how we might best model attention and consciousness in four-steps, the first two in this section being about attention proper, and the further two steps in the following section as to how consciousness might be created by extending our model of attention. In such a manner it is hoped to make the basis of CODAM clearer than approaching it in only one big step.

6.1.1.1 Step 1

In step 1, we specify attention as a control system. In other words there are considered to be two parts of the brain that can be differentiated from each other:

1. The Controlled portion of the brain;
2. The Controller portion of the brain.

The attention control system works in the obvious manner of sending control signals from the controller to the controlled parts of the brain to achieve a desired end or goal:

$$\text{Controller} \rightarrow \text{Controlled}$$

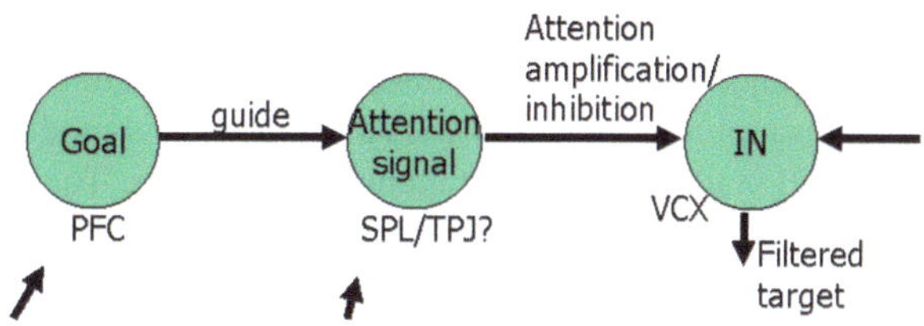

Fig. 6.1 The ballistic control model of attention

This bipartite division of the brain has been justified by many brain experiments, on a range of mammals from humans to rats; in the human case see for example (Bressler et al. 2008; Corbetta and Shulman 2002). The controlled portion of the brain consists mainly of the lower-level sensory areas involved in processing input features, like the visual or auditory cortices. In the former, for example, there is a breakdown of an input stimulus into feature components of ever increasing complexity along the hierarchy of V1, V2, V3, V4 and so on, concluding with object representations in the temporal lobe. Each of these modules can be attended to, in a correlated fashion, when attention is paid to a visual stimulus. On the other hand the controller consists of parietal and prefrontal regions involved in setting and holding goals (of where and/or what to attend to) and in sending an attention signal back to the controlled regions in order to amplify the representation of the attended stimulus in the visual (or other sensory) field.

Again much experimental data supports this notion that the attention signal from the controller to the controlled region amplifies the activity of those neurons involved in representing the attended stimulus whilst at the same time inhibiting those neurons involved in distracter representations. Some of this evidence has been outlined in the previous chapter (as in the results of Bressler et al. 2008). It is in this way that the brain representation of the attended stimulus becomes dominant in the sensory cortices and can be used for higher-level processing, such as thinking, reasoning etc.

The above controller/controlled division of the brain is not all that attention does, since it is even possible that it can fold back and 'attend to itself', but such further extensions do not change the main work that attention performs for manipulating inputs in the brain from the outside world.

6.1.1.2 Step 2

Step 2 can now be taken, by separating the controller brain region into a biasing or Goal module and a module for creating the attention signal to move the focus of attention under the guidance of the Goal module. These two modules are shown in the Ballistic Control model of attention movement in Fig. 6.1, with the goal module (denoted 'Goal' in Fig. 6.1) being in the prefrontal cortex (PFC) and the generator of the signal causing movement of the focus of attention (denoted 'Attention signal' in Fig. 6.1) being in the superior parietal lobe (SPL) or in the temporo-parietal junction

(TPJ). The third module in Fig. 6.1 is denoted ‘IN’, being the visual cortex (VCX), where object and spatial features of the input stimulus are extracted in the cortex. Various inputs to the Goals and Attention Signal modules are shown, corresponding to emotional and long-term memory influences. There is also the important output from the IN module, being the attention-amplified target representation and denoted ‘Filtered Target’ in the figure. It is this latter signal which the faculty of attention delivers to the rest of the brain. Such an architecture has numerous experimental supports, especially the results of (Bressler et al. 2008) already referred to and discussed in the previous chapter.

In the figure the sites observed by brain imaging techniques activated by attention tasks are written beneath each of the functional modules. PFC = prefrontal cortex; SPL = superior parietal lobe; VCX = visual cortex; TPJ = temporo parietal junction.

6.1.2 Steps 3 and 4: Extending Attention to Consciousness

6.1.2.1 Step 3

Ballistic control occurs in the simplest example of firing a gun – it is aimed and then fired. Once fired it is not possible, by feedback, to modify where the bullet is going. A step towards improving such control, being step 3, is to introduce a site where the attended stimulus can be stored for a short time (such as over a few seconds). Usually called a buffer or working memory site, such sites have also been observed being able to hold activity for a few seconds (Jonides et al. 1998; Ravizza et al. 2005; Husain et al. 2001); we considered such sites in detail in the previous chapter. This buffered activity could be used to correct the mechanism to produce it if the buffered activity were in error, as compared to the goal stimulus to be attended to. We will come back to the error-correcting process shortly.

It is now accepted that buffer sites (there are more than one, these being different for different modalities) in the brain play an important role in consciousness. If they are destroyed by a stroke then the associated conscious experience is annihilated. That experience will return if there is recovery of these buffer sites. It has been accepted in neuroscience that these buffer sites provide the content of the conscious experience of the particular attended stimulus (in association with lower level stimulus activity in the various feature modules). It is thought that a given attended stimulus attains awareness due to its activity being augmented by attention, with distracters being rejected (by inhibition from the attention feedback signal). Such an acceptance/rejection process could occur by access to the buffer site being through incoming activity, amplified by attention, attaining a threshold of activity, and thereby activating the relevant buffer code. This helps explain why attention is needed for consciousness of a stimulus.

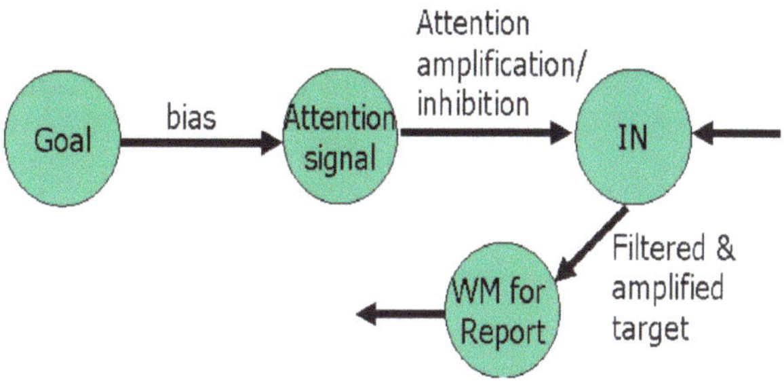

Fig. 6.2 The extended ballistic attention control model, with activity on the extra working memory buffer (beyond the modules in Fig. 6.1) allowing for further use of the representation of the attended stimulus, for reasoning, thinking, etc by the subject

The extended ballistic model of attention control is shown in Fig. 6.2. The working memory module (denoted 'WM for Report' in Fig. 6.2) is used to store the attended stimulus representation for report to other such sites, so leading to thinking or reasoning, or just reporting.

The extended ballistic control model, from what was said above, would appear to possess room for explaining consciousness, at least of the content of stimuli in the environment being attended to. But there is the important question: who has the experience of that consciousness? There would appear to be no inner self that can have such an experience. Just holding activity in a brain site does not mean that activity becomes part of the conscious experience of the owner of that brain. Continued activity can occur in various sites, such as hippocampus, but that organ can be lost without loss of consciousness, as in the well known case of the subject HM (whose hippocampus and middle temporal lobes were removed to reduce his terrible epileptic seizures, but who remained conscious till the recent end of his life).

Something is still missing in the attention model created so far. Who is having the conscious experience? We must attempt to discover how such experience might arise from a more complex model of attention, since otherwise we will only be able to explain the experiences of a zombie – which are completely absent.

6.1.2.2 Step 4

The fourth and final step to a model of consciousness is that taken in CODAM: this involves the presence and use of a corollary discharge (a copy) of the attention movement signal, so as to provide greater efficiency in moving the focus of attention. It is this final step that is more conjectural but needs to be made to bring attention control into the modern age. The resulting architecture is shown in Fig. 6.3. Here the goal module, providing the bias to where the attention focus must move, is just indicated by the bias input on the left. This causes the 'Attention' module to generate the required movement of the focus of attention to the relevant biased position or object and so to amplify the input activity representing the attended stimulus (spatially or as a specific object). The IN module denotes the visual cortex for visual attention, as before.

Fig. 6.3 The CODAM model

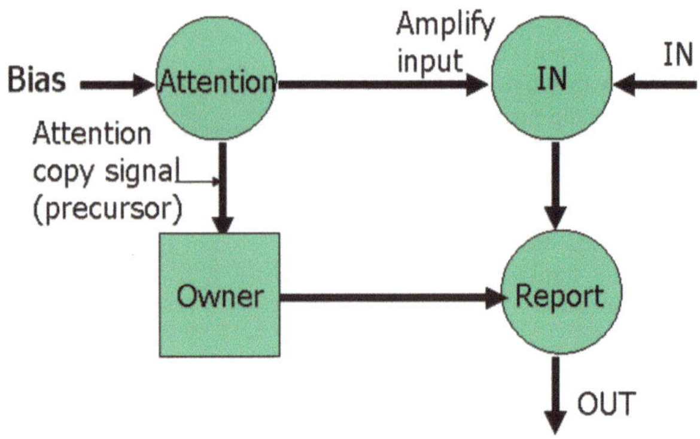

The Attention copy signal is shown proceeding from the 'Attention' module to a module denoted 'Owner' in Fig. 6.3. This acts as a buffer to hold the content of this attention copy signal for a short time. The buffered signal will thereby be available for a number of things: to speed up the access of the attended stimulus activity into the working memory module and also to correct any errors that might be made (such as allowing distracters to creep into the working memory module and so into the content of consciousness). The copy signal can also be used to increase the posterior-going attention signal if there is a possible problem with a distracter. This extension thereby allows corrections to be made to attention movement, in comparison with the inflexible ballistic control. It is important to notice that these corrections can be made early in the creation of consciousness of a given stimulus, since the corollary discharge is available immediately the attention movement signal has been produced, and does not have to wait on the amplification of posterior neural activity (which may take some 100–200 ms).

It is the attention copy signal which, it is claimed in CODAM, provides the 'owner' content of the relevant neural activity. It generates the experience of the 'inner self', that of ipseity. As such the CODAM extension of Fig. 6.3 allows for the inclusion of the necessary complexity of consciousness creation by attention to begin to tackle, for example, the experiences of schizophrenics.

The various modules are explained in the text by the figure. The only new module is the 'Owner' module not present in Figs. 6.1 or 6.2.

6.1.3 *Evidence for CODAM*

What is the evidence for the existence of such an attention copy signal in the brain, and hence of a CODAM-style of attention control architecture? There are several such lines of evidence, of which the three main ones are (1) the proposed premotor model of attention (Rizzolatti et al. 1994) (2) the improvement of control achieved by the use of an attention copy signal (3) experimental results coming from the attentional blink (Sergent et al. 2005), trying to model these results through CODAM (Fragopanagos et al. 2005), and how the attentional blink is increased in schizophrenics, with again modeling of experimental results by CODAM (Wynn et al. 2006).

1. The premotor model of attention (Rizzolatti et al. 1994) is based on the idea that attention and motor control share a certain amount of brain circuitry at the higher levels in the brain. A copy or corollary discharge model has been used by a number of researchers to better understand features of motor control in the brain (Desmurget and Grafton 2000), and a motor control copy or corollary discharge signal even been directly observed there (Diamond et al. 2000; Sommer and Wurtz 2002). The attention copy model of CODAM is a natural extension of the motor control processes in the brain to its companion faculty of attention. The considerable overlap of brain sites, such as in eye movement control versus covert (eyes fixed) attention control would lead one to expect a similar corollary discharge to exist for attention control as it does for motor control. CODAM fulfils that expectation.
2. The range of models of engineering control was enormously enlarged over the past decades by the development of models employing a corollary discharge of the control signal to make the control more efficient. This was achieved by inserting into the control structure a predictor or forward model of the expected state of the controlled system where, for example, the state of a steel-making plant would at its simplest be the temperature of the steel, and its forward model would predict the change in temperature of the steel when a certain amount of coal was used; for attention in the brain the state of the controlled system is that for the neuron activity of the 'Report' buffer of Fig. 6.3, not for the whole environment in lower sensory brain areas, while the predictor would be the corollary discharge buffer, so the activity on the 'Owner' module of Fig. 6.3. Such a predictor uses the corollary discharge signal to make a prediction of what the next state of the system would be if the control action were to be used on the system (such as feeding heat to the steel-making plant). Such a prediction would then be used to check for (and thereby correct) errors about to be made by the original control action, as described earlier.
3. The process of filtering out the distracters, leaving solely the attended stimulus activity for further manipulation as in reasoning, thinking, etc, is an important component of attention. With the extended attention copy model of Fig. 6.3 the owner module is the predictor, and its output will thereby be the predicted state of the attended stimulus (ahead of its arrival). This predictor activity can be used to compare with that on the goal module. The difference between these two neural activities will be an error that can be used to modify the attention control signal accordingly (to remove the error) even before arrival of the attended stimulus to the working memory buffer. The output of the owner module can also be used to speed up access to the working memory report module of Fig. 6.3. In both ways (error correction and speed-up of attention amplification) greater efficiency will thereby be obtained in attention control. Evolutionary theory would lead us to

expect such mechanisms therefore to be used in the brain. This leads to a clear function for consciousness (as seen in the CODAM model): to make the bringing of stimuli into the focus of attention more efficient (faster and more error free). This negates the idea that consciousness is a pure epiphenomenon.

4. The attentional blink (AB) has been much studied in normal people over the last few decades due to the paradigm being able to explore how attention can be broken down by a very difficult task and thereby several of its components exposed. The AB involves a rapid serial presentation of visual stimuli (RSVP denoting 'rapid serial visual presentation'), such as digits and letters, at a rate of about 10 Hz. In the paradigm used recently (Wynn et al. 2006) to compare the AB between schizophrenics and normal controls, the RSVP stream involved the digits 2, 3, 4, 5, 6, 7, 8, 9 as distracters and the capital letters A, C, E, J, K, R, T and Y as targets. The task for both classes of subjects (37 schizophrenics and 26 normal controls) was to correctly identify a first target (T1) (at the fixed focus of attention) when it appeared, and then to identify a second target (T2) (also at the same attention focus) after the appearance of the first. The second target could appear at one of several lags after the first, thus testing the difficulty of such identification by determining the probability of determining T2 given T1 was correctly detected, denoted by Pr(T2|T1), for various lags. The resulting values of Pr(T2|T1) were then plotted against the time lag of the detection of T2 after T1 (from Lags 1 to 5 in the first instance).

In normal people the AB arises as follows. Some time is expected to be taken by a subject attending to processing the first target T1. If the second target T2 is shown too soon after the first then T2 may not be allowed to be processed by the events involved in the processing of T1. Thus if T2 is presented within a few lags after T1 there is observed a dip in the plot of Pr(T2|T1) against lag number for control subjects, with the least value of Pr(T2|T1) being at about 3 lags (or 270 ms) after T1. That time delay is thought to occur when attention is being focused maximally on T1, it has been suggested, so preventing attention being spread further to T2 (Shapiro et al. 1994): it is when 'attention blinks shut'.

Furthermore, it was found (Wynn et al. 2006) that the curve for the values of Pr(T2|T1) for schizophrenic patients over the set of 5 lags was very similar to that for the normal controls but was lowered by a constant value of about 0.2 at each lag. The minimum of the U-shape of the two curves was similar (this being the attentional blink itself) lasting from lag 2 to lag 4, and a minimum at lag 3. A check was made that neither of the results was due to visual masking by increasing the duration of the distracter stimuli after T1 or T2. This modification did not increase the difference between the normal controls and the schizophrenics, so implying that there was little effect of visual masking and the main effect was due to the difficulty that T2 experienced in trying to gain access to an appropriate visual short term site (called the 'WM for Report' site in Fig. 6.2) while T1 was still being processed. Thus the attentional blink is larger for schizophrenics as compared to normal controls, due, it is expected, to schizophrenics having less control over their movement of attention focus (from T1 to T2 in this case).

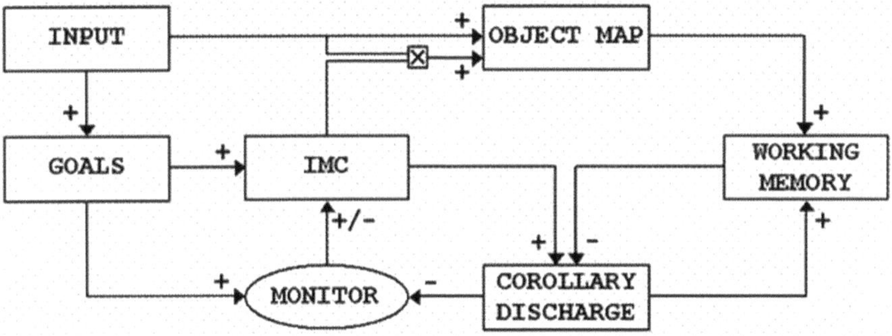

Fig. 6.4 The full CODAM model of the control of attention (Taylor 2000, 2007)

A detailed mathematical formulation of the CODAM model, with the more complete architecture of Fig. 6.4 (to be discussed shortly), was used to simulate the AB (Fragopanagos et al. 2005) and showed results consistent with numerous publications of the U-shape of the curve of Pr(T2|T1) against the time lag of T2 after T1. The basic mechanism of the CODAM model leading to this result on the AB in the Rapid Serial Visual Presentation case was due to the inhibition present in the detailed CODAM model being used. This arose from the activity of the first target achieving its access to the 'WM for Report' module of Fig. 6.3 also inhibiting distracter-based activity trying also to gain access to that module. Since T2 will be regarded by a subject's brain as a distracter compared to T1 (at least if T1 is still being processed) then there will be inhibition of T2 so bringing about the damage to its detection as shown by the attentional blink (especially damage to the corollary discharge signal associated with T2 on its 'Owner' buffer (Vogel et al. 1998)).

Furthermore, the model results were supported by more recent data of Sergent et al. (2005) on the presence of inhibition of T1 on its Report module caused by the activity of the attention copy model for moving the focus of attention to T2. The resulting alterations of the usual sequence of brain waves observed during conscious report (so access to the 'WM for Report' module of Fig. 6.3) fitted very well with the simulation results of these inhibitory processes obtained from an extended version of CODAM with inhibitory feedback from the 'Owner' module of Fig. 6.3 to distracter nodes on the 'WM for Report' module of Fig. 6.3 (Taylor and Fragopanagos 2007).

Finally, simulations were also made of how damage to the input of the WM Report module of Figs. 6.3 or 6.4 (by reducing the strength of the response for each neuron in the modules feeding to it, especially the corollary discharge buffer or 'Owner' module of Fig. 6.3) can affect the AB curve (of Pr(T2|T1) against the time lag between T1 and T2) so as to mimic what might be occurring in the case of a schizophrenic. The simulation results (Fragopanagos and Taylor, 2006, personal communication) fitted closely the modification of the AB curve for schizophrenics as compared to that for controls in (Wynn et al. 2006), in that there was a reduction of the ability to observe T2 due to the reduced T2 input to its buffer, the WM Report module of Fig. 6.3, given that T1 was observed. Reduction of such input could be due to the reduction of output from the 'Owner' module of Fig. 6.3, consistent with a reduction of the sense of ipseity in schizophrenics.

We will present in Chap. 9 a more experimentally-based detailed justification of the existence of the corollary discharge of attention movement.

6.2 Overall Conclusions and the Full CODAM

The basic postulate of the attention copy approach to consciousness is that the experience of ownership of a phenomenological experience is generated by the activity of this attention copy signal residing briefly on its component of the relevant working memory site. Such a site is shown in Fig. 6.4 as the 'Corollary Discharge' module (and equal to the 'Owner' module of Fig. 6.3). The architecture of Fig. 6.4 is a more complete (and complex) extension of that of Fig. 6.3 and was that used in the simulation of the attentional blink and other attention-based paradigms (Fragopanagos and Taylor, personal communication, 2006; Fragopanagos et al. 2005; Taylor and Fragopanagos 2007; Korsten et al. 2006). There is now included in Fig. 6.4 the Goals module (biasing the movement of the focus of attention, as in Fig. 6.2), the corollary discharge module (the 'owner' module of Fig. 6.3) and the new 'Monitor' module (used to generate the error signal between the predicted future attended state of the visual cortex, as contained in the corollary discharge module, and the Goals module). The module generating the signal to move the focus of attention, so at the centre of attention control, is denoted IMC (for inverse model controller) and is identical to the 'attention' module of Fig. 6.3. Finally the input module IN of Fig. 6.3 has been extended to both an 'input' module and an 'object map' module in Fig. 6.4. The arrows indicate the direction of flow of information, with the positive or negative signs indicating if the input to a given module is excitatory or inhibitory. The extra cross symbol in Fig. 6.4 on the input from the inverse model controller to the object map indicates that the attention feedback from the IMC is suitably fused with the input to the object map (in the original simulation a multiplicative fusion was used).

The action of the various modules of Fig. 6.4 and the relation to those in Fig. 6.3, are explained in the text.

We note that there is a crucial difference between the CODAM attention control architecture of Fig. 6.3 (or its more complete form in Fig. 6.4) and that for standard control theory or that suggested as arising for motor control in the brain (Desmurget and Grafton 2000): the attended state estimator (denoted 'working memory' in Fig. 6.4) and the predictor module (denoted 'corollary discharge' in Fig. 6.4) only concern themselves with the attended target stimulus and not with distracters (except to inhibit them). Thus there is not an estimate of the whole environment as contained in the posterior cortex (a role played by posterior visual cortex in the case of vision), but only of the attended state of the environment, where the environment is represented in the brain by the activity across lower level posterior cortex. This is the basis of the filter process of attention, and produces a working memory module activation in which distracters are not represented; this is a crucial step in a complex environment to reduce its complexity for efficient higher level processing.

To repeat, improvement in attention processing (so going from step 3 to step 4 above) is achieved by the use of the attention copy signal buffered for a short time on the corollary discharge module. As noted in (Sergent et al. 2005) for the attentional blink, there is inhibition of the access to the 'working memory' module in Fig. 6.4 for the first target by the access of the second to its 'corollary discharge' module of Fig. 6.4, as they report in their experiments (Sergent et al. 2005). It is that sharpening-up process which has been suggested to be carried out by the attention copy signal, as a process of filtering out distracters (and is described by inhibition in the input from the corollary discharge module of Fig. 6.4 to the working memory site, needing to be added to the excitation sign of Fig. 6.4). At the same time the monitor can be used to re-activate the attention signal (generated by the attention controller) by a rehearsal goal (Korsten et al. 2006). Thus the various components of the extended architecture of CODAM can achieve considerable processing efficiency by means of the several features of attention control the model contains. As such, the CODAM architecture is an attention analogue to the motor control models proposed as making motor control more efficient than by use of a purely ballistic control system, although with added inhibition (from the corollary discharge signal) to prevent distracter access to the input buffer, and the filtering process to leave only the attended stimulus activity as the attended state representation on its working memory buffer.

The 'immunity to error by misidentification of the first person pronoun' feature (Shoemaker 1968) is an important component of experience. It arises from the fact that one cannot ask a friend, when they say they are in pain: 'Are you sure it is you who are in pain – could it be someone else?' They are sure it is they who are in pain. This feature of experience can be explained as part of CODAM by the lack of error in the boosting process of the attention feedback signal (together with the inhibition of distracters). If an incorrect stimulus (a distracter) is trying to enter the visual working memory then it will be inhibited by the ownership signal (from activity on the 'corollary discharge' module of Fig. 6.4), which will only let through to awareness of content that target stimulus representation which is desired by the original goal (for endogenous attention) or has grabbed attention control (for exogenous attention).

We note that originally this feature of experience (immunity to error about 'I') was claimed to have logical status (Shoemaker 1968), but more recently experiments have shown that the feature has only empirical grounds for validity. Thus in (Mizumoto and Ishikawa 2005) it was shown by the use of a head-mounted display (showing the user an externally based view of his body) that such immunity could be lost. A similar situation could arise in a CODAM-based version of the inner self, where breakdown of parts of the system could destroy the properties required to give a unique attribution to oneself of incoming stimuli onto the relevant working memory buffer. We will discuss an example of that in the case of hallucinations in schizophrenia later.

The copy of the attention movement signal held on the corollary discharge buffer of Fig. 6.4 is proposed as the source of the inner self. The related neural activity will function, it has been proposed (Taylor 2000, 2007, 2010a, b) so as to represent the owner of the about-to-be-expected attention-amplified activity due to arrive at the sensory working memory module of Fig. 6.4. Such an interpretation of the source of the ownership of conscious content is consistent with the interpretation of the inner self as being content free (Zahavi 2005). This is due to the high level coding of activity in the corollary discharge module of Fig. 6.4 and that in the attention controller or IMC of Fig. 6.4 (from which it receives its activity). There is other neural activity to which the corollary discharge module has access, such as that of the Monitor and the attention controller, besides that on the sensory working memory buffer, all providing further expansion of the ownership activity on the corollary discharge buffer. These other activities involve crucial processing stages, so that the corollary discharge module functions as being at the centre of a network of highest-level processing sites in the attention network. This module therefore has high-level content which preserves its character of 'nothingness' (Sartre 1943) but expands its processing powers.

We add that nothingness was an important concept in Sartre's philosophy (Sartre 1943), and as noted in (Sokolowski 2000, p. 220), in a comment on Sartre's philosophy: "; negation is not merely a feature of our judgments, but is given in the intuitive experience that precedes judgment." In CODAM this intuitive experience of negation preceding judgment is created by the activity of the corollary discharge just before the access of the content of consciousness, describing a particular attended stimulus, to attain its buffer working memory, as explained earlier.

Due to the importance of this step into the unknown of including a corollary discharge of the attention movement control signal just carried out, let me repeat again some of the arguments for it. It was first suggested in (Taylor 1999, 2000) that a corollary discharge of the attention control signal from the IPS/SPL complex is sent to another component of the parietal region to act as a forward model and hence speed up attention movement itself (possibly related to the effects observed through TMS by Chambers et al. 2004a, b just mentioned). This discharge signal explains the origin of the acronym CODAM as COrollary Discharge of Attention Movement. Such additional architecture would lead to an early error signal, before the arrival in the working memory site of the amplified signal representing the attended stimulus. Thus speeded attention movement, along with its being able to be corrected early in the processing if an error in the attention movement is predicted, leads to a more efficient attention control system than provided solely by a ballistic control model (with no feedback).

The second reason for extending the ballistic control network for attention to a CODAM style has already been discussed, in the cases of modelling the attentional blink AB (Fragopanagos et al. 2005) and other similar paradigms (Fragopanagos and Taylor 2007, private communication). It also relates to experimental data for the AB observed as apparent inhibition of the site of the posterior P3 for a first target by activity at the N2 timing of a second target (Sergent et al. 2005). This is implemented in the CODAM model by inhibition of distracters from the corollary discharge signal being sent to the relevant working memory site.

Finally, the third reason is the premotor theory of attention (Rizolatti and Craighero 2010). Consider the great impact of modern control theory concepts on motor control by the brain (Kawato 1999; Desmurget and Grafton 2000) and of mental ill health, where much use has been made of a corollary discharge of the motor control signal. This has led to the search and final proof of the existence of such a motor control corollary discharge signal for eye movement (Wurtz and Sommers 2004). A similar extension of attention control from ballistic to involving a corollary discharge is thus expected. For if the brain evolved to use a corollary discharge signal in its motor control processes it is to be expected that similar evolutionary pressures led to the evolution of a corollary discharge signal for attention control.

There is still the question as to how the DAN/VAN architecture of Fig. 5.1 might be extended to take account of a suitable corollary discharge, assuming that signal exists. Such a signal arises as a copy of a basic control signal generated by an IMC purely to control the parameters of a plant. In the case of attention in the brain, there may be two such IMCs, as discussed in the previous section: that in the DAN (IPS/SPL) and that in the VAN (TPJ). The second of these sites is left free to be active as a circuit-breaker in the case of inputs that possess a feature carried by a target in a search paradigm. But we are now in danger of having a plethora of corollary discharge signals to consider. Let us proceed with caution in their analysis.

There is still the question of reconciling the architecture of Fig. 5.1, composed of what appear to be two separate ballistic attention control networks, with that of Fig. 6.4, which has only one such overall network. It is possible to take two networks of architecture of the form of Fig. 6.4, so as to duplicate the extension of the single ballistic attention control network to the DAN and VAN dorsal and ventral networks of Fig. 6.4.

Whilst these results are indicative of greater things to come the present level of results cannot help to move the CODAM model forward. What is needed is proposals and later proof of the presence and positions of the extra modules mentioned above concerned with the CODAM model.

6.3 The Corollary Discharge Component

The siting of all the modules in the CODAM model of Fig. 6.4, as discussed in the previous sections, is in agreement with that assumed in the first formulation of CODAM (Taylor 1999, 2000) and in its later developments and applications (Taylor 2007). The error monitor may be both sited in the cingulate as well as more locally distributed. Thus all but one component of the engineering control based model CODAM of attention have been supported and sited in various parts of the brain. However there is still the question of the crucial component of CODAM – that of the corollary discharge. Does this signal exist and if so where should we look for it in the brain? A similar situation arose in the motor control approach to motor activity by the brain, and has been analysed in several papers (Desmurget and Grafton 2000; Wolpert and Ghahramani 2000). As noted earlier, direct evidence of such activity in the brain for eye movement control has been presented by (Wurtz and Sommers 2004). A similar situation was earlier suggested as having occurred in the experiment of (Sergent et al. 2005) on the attentional blink. We naturally ask if any of the experimental data in the above references, other than that of (Sergent et al. 2005), are relevant to this search for the corollary discharge of the attention movement control signal? We may also ask if the engineering control approach used to develop CODAM can also be used to make more precise the model presented in (Postle 2006)?

The claims in (Postle 2006) mentioned in Chap. 5 fit very well into the CODAM model. To repeat, Postle proposed that the slave buffer sites are based in posterior cortical sites and not in the prefrontal cortex. However he suggested that slave sites are in those places where the highest level coding of input stimuli occurs. Thus the working memory module has to be modified to take account of the feature that the distinction between the object features and the working memory module may be more blurred than represented by the CODAM model in Fig. 6.1; in any case that is part of the more general blurring of the functionality of all the modules in Fig. 6.4, since there is much more overlap of function in brain modules than as contained in distinctions of functionality in the modules of Fig. 6.4. However that modification still leaves the corollary discharge signal and its associated buffer site unassigned (modulo the data of Sergent et al. 2005).

One important function that is claimed to be carried out by the corollary discharge signal in CODAM is its early amplification of the input working memory module, and its inhibition of distracters. But such an amplification process itself has been carefully studied in its own right by Nobre and colleagues by means of retro cueing (Lepsien and Nobre 2006, 2007). In this process, attention is directed specifically onto the input working memory activity already set up, so as to achieve the amplification of such activity (after initial stimulus encoding in working memory). Such retro-cueing leads to the question: to which parts of the brain can attention be directed? Is it all parts of the brain or only certain proscribed regions? The CODAM model implies that it is directed initially at what is known in engineering control terms as 'the plant'. For CODAM the plant consists of lower-level cortices where feature analysis is performed on inputs. Indeed that appears to be appropriate, when considered from the viewpoint of early processing analysis. However there is uncertainty about where the higher-level object representations fit in such an attention-controlled scheme.

Are these higher-level representation part of the processing analysis or do they sit above that? The reaction times in the retro-cueing paradigm of about 600–800 ms, with advantage of the retro-cues of about 80–150 ms, depending on working memory load, were as measured by (Lepsien and Nobre 2006, 2007). These values were thus of the same order as for pre-cueing. The representations being attended to would thus be expected to be classified also as 'plant': the attention retro-cueing needs as much apparatus and time to complete as does the pre-cued plant material in the brain. In other words the corollary discharge seems to be as much bound up (and hidden) in the retro-cueing as it does in the pre-cueing case. The conclusion of this is that retro-cueing does not provide a 'naked' view of the corollary discharge, only a view of the same character as for other attention processing. The corollary discharge signal remains as elusive as ever.

It is natural to ask if there are any sites in the brain to which attention cannot be directed? The functions of the goals coded in PFC certainly can, although the process of such attention would be expected to create activity on buffer sites which correspond to a particular goal and lead to the whole panoply of the CODAM-style attention control system being employed. Thus we expect that all sites in the brain are accessible to attention control, but only through the overall attention control systems described so far. As such then, they provide no new avenue of attack on the corollary discharge signal at the heart of the CODAM model.

6.4 Conclusions

Attention is shown to be effectively described by engineering control methods, as discussed above. It involves an attention control signal generated by a specific part of the brain, mainly in the parietal lobes. It is assumed that this signal arises by competition between various biasing signals in a complex environment. Thus the attention signal choice is biased by prefrontal-based goal signals. Such a ballistic attention control structure is valid for both endogenous (top-down) and exogenous (bottom-up) attention. In the former case of endogenous attention these bias signals arise from prior goals coded on-line by continued prefrontal cortex activity; in the latter case of exogenous attention the bias signals are generated afresh in prefrontal cortex by external stimuli trying to capture attention.

This simple control structure needs to be extended to include

(a) an error monitor, based in cingulate (and possibly insula and part of prefrontal cortex, especially inferior);
(b) a slave working memory buffer or slave site, comprising brain regions coding for object or spatial representations (FFA, PPA and PPC for faces, scenes and spatial positions respectively). These latter sites are proposed to provide information as to the content of awareness. They involve activity able to be used in more general areas of the brain for cognitive processes such as thinking, reasoning, planning etc. These slave sites can also be put under attention control, as by retro-cueing;
(c) A further extension has been proposed in the CODAM model of attention (Taylor 2000, 2007). A copy of the attention control signal is assumed to be created by the attention movement signal generator. This copy (the corollary discharge signal) is assumed to be able to be held on a suitable buffer site

to enable various manipulations being made in the ballistic attention control model to occur more efficiently. Thus attention amplification of the input signal (representing the attended stimulus), to help this signal to gain access to the buffer site, is speeded up by pre-amplification by the corollary discharge signal directed at the buffer site of the input signal.

This structure is powerful enough, it has been claimed (Taylor 2010a, b) to be able to explain the creation of the inner self and hence of the most crucial component of consciousness. Without an inner self no creation of consciousness would appear possible, for there would be 'no-one' to have the conscious experience. Support for the existence of such a corollary discharge signal arises from the possibility of using it, and the associated damage to the sense of self when the corollary discharge buffer is itself damaged, to help explain the symptoms of schizophrenia (Taylor 2010a, b).

Later (in Chap. 9) more detailed experimental evidence will be presented for the existence of the corollary discharge signal. However before that we will consider in Chap. 7 the manner in which the inner self is fully represented by this corollary discharge, and subsequently raise a serious question about the existence of the inner self and related phenomena in Chap. 8.

References

Bressler SL, Tang W, Sylvester CM, Shulman GL, Corbetta M (2008) Top-down control of human visual cortex by frontal and parietal cortex in anticipatory visual spatial attention. J Neurosci 28(40):10056–10061

Chambers CD, Stokes MG, Mattingley JB (2004a) Modality-specific control of strategic spatial attention in parietal cortex. Neuron 44(6):925–930

Chambers CD, Payne JM, Stokes MG, Mattingley JB (2004b) Fast and slow parietal pathways mediate spatial attention. Nat Neurosci 7(3):217–218

Corbetta M, Shulman GL (2002) Control of goal-directed and stimulus-driven attention in the brain. Nat Rev Neurosci 3:201–215

Desmurget M, Grafton S (2000) Forward modelling allows feedback control for fast reaching movements. Trends Cogn Sci 4:423–431

Diamond MR, Ross J, Morrone MC (2000) Extraretinal control of saccadic suppression. J Neurosci 20:3449–3455

Fragopanagos N, Kockelkoren S, Taylor JG (2005) A neurodynamic model of the attentional blink. Cogn Brain Res 24:568–586

Husain M, Mannan S, Hodgson T, Wojciulik E, Driver J, Kennard C (2001) Impaired spatial working memory across saccades contributes to abnormal search in parietal neglect. Brain 124:941–952

Jonides J, Schumacher EH, Smith EE et al (1998) The role of parietal cortex in verbal working memory. J Neurosci 18:5026–5034

Kawato M (1999) Internal models for motor control and trajectory formation. Curr Opin Neurobiol 9(6):718–727

Korsten N, Fragopanagos N, Hartley M, Taylor N, Taylor JG (2006) Attention as a controller. Neural Netw 19(9):1408–1421

Lepsien J, Nobre AC (2006) Cognitive control of attention in the human brain: insights from orienting attention to mental representations. Brain Res 1105(1):20–31

Lepsien J, Nobre AC (2007) Attentional modulation of object representations in working memory. Cereb Cortex 17(9):2072–2083
Mizumoto M, Ishikawa M (2005) Immunity to error through misidentification and the bodily illusion experiment. J Conscious Stud 12:3–19
Postle BR (2006) Working memory as an emergent property of the mind and brain. Neuroscience 139:23–38
Ravizza SM, Behrmann M, Fiez JA (2005) Right parietal contributions to verbal working memory: spatial or executive? Neuropsychologia 43:2057–2067
Rizolatti G, Craighero L (2010) The premotor theory of attention. Scholarpedia 5(1):6311
Rizzolatti G, Riggio L, Sheliga BM (1994) Space and selective attention. In: Umilta C, Moscovitch M (eds) Attention and performance XV. Conscious and unconscious information processing. MIT Press, Cambridge, MA, pp 231–265
Sartre J-P (1943) Being and nothingness. Routledge, London
Sergent C, Baillet S, Dehaene S (2005) Timing of the brain events underlying access to consciousness during the attentional blink. Nat Neurosci 8:1391–1400
Shapiro KL, Raymond JE, Ansell KM (1994) Attention to visual pattern information produces the attentional blink in rapid serial visual presentation. J Exp Psychol Hum Percept Perform 20:357–371
Shoemaker S (1968) Self-reference and self-awareness. J Philos 65:556–570
Sokolowski R (2000) Introduction to phenomenology. Cambridge University Press, Cambridge
Sommer MA, Wurtz RH (2002) A pathway in primate brain for internal monitoring of movements. Science 296:1480–1482
Taylor JG (1999) Race for consciousness. Bradford Book, Cambridge, MA
Taylor JG (2000) Attentional movement: the control basis for consciousness. Soc Neurosci Abstr 26:2231#839.3
Taylor JG (2007) CODAM: a model of attention leading to the creation of consciousness. Scholarpedia 2(11):1598
Taylor JG (2010a) A neural model of the loss of self in schizophrenia. Schizophr Bull 37(6):1229–1247
Taylor JG (2010b) The I's eye view of consciousness. J Conscious Stud 17(1–2):95–117
Taylor JG, Fragopanagos N (2007) Resolving some confusions over attention and consciousness. Neural Netw 20(9):993–1003
Taylor JG, Cleeremans A, Freeman WJ (2007) Brain and consciousness. Neural Netw 20:929–1060
Vogel EK, Luck SJ, Shapiro K (1998) Electrophysiological evidence for a post-perceptual locus of suppression during the attentional blink. J Exp Psychol 241:1656–1674
Wolpert DM, Ghahramani Z (2000) Computational principles of movement neuroscience. Nat Neurosci 3:1212–1217
Wurtz RH, Sommer MA (2004) Identifying corollary discharges for movement in the primate brain. Prog Brain Res 144:47–60
Wynn JK, Breitmeyer B, Nuechterlein KH, Green MF (2006) Exploring the short term visual store in schizophrenia using the attentional blink. J Psychiatric Res 40:599–605
Zahavi D (2005) Subjectivity & selfhood. MIT Press, Cambridge, MA

Chapter 7
The Owner, The Inner Self and CODAM

7.1 Inside CODAM

The functioning of the pre-reflective or inner self is considered in more detail in this chapter in terms of its possible creation through the CODAM model of attention presented in the previous chapter. In contradiction to the view of Western phenomenology, that the inner self appears to serve no specific purpose except that of providing the further ownership of experience, it is proposed here that the inner self acts rather as a call centre, enabling connections to be made between distant and functionally different components of brain processing thereby making such interactions more efficient. It achieves these functions by monitoring and speeding up ongoing calls or those calls about to be set up so that incorrect call routing is avoided and call switching occurs as fast as possible.

Consciousness is still a puzzle. It used to be a philosophical puzzle, but now it has also turned into a scientific one. The main puzzle is that of how inanimate matter could be so configured as to lead to the matter thinking and reasoning and above all experiencing all of these ongoing processes. It is this latter puzzle, above all – 'how can there be experiencing matter?' – which has led to much controversy. In this chapter we develop a solution which has both scientific potential (it can be tested scientifically) as well as consistency with the inner nature of our experience, as we live it from day to day.

These twin aspects – being both scientific and based on personal experience – seem to be so diametrically opposed to each other that they appear totally mutually incompatible. As is usually said, science is based on an objective view of reality – always from the outside. It takes only a third person perspective on events, never a first person one. On the other hand inner experience, by its very nature, is obtained from a subjective view of the universe. It is only a first person view of the world. How can these two apparently mutually opposing views be reconciled?

To achieve the desired reconciliation we will start from the scientific viewpoint. That will allow us to build on a sure foundation: if the theories and models

J.G. Taylor, *Solving the Mind-Body Problem by the CODAM Neural Model of Consciousness?*, Springer Series in Cognitive and Neural Systems 9,
DOI 10.1007/978-94-007-7645-6_7,

we construct are well founded scientifically then at least the foundations of our approach will not fail, nor lead us into absurdities and falsehoods.

From our firm scientific base we must then attempt to build a bridge to inner experiences that the matter our theories and models are dealing with might possess. Such bridge building is clearly difficult but there are various stepping stones we can use to cross over. One crucial stepping stone is the report from subjects as to their inner experiences and the relation of that to particular forms of activity in the matter being considered.

It is clear from the above paragraph that we can only be talking about matter in the form it is found in the brains of animals thought to possess consciousness. To avoid much further controversy we will restrict ourselves here solely to human brains. This is not too restrictive since there are about seven billion such brains to take data from on their inner experiences.

7.2 The Existence of the Inner Self, the 'I'

The existence of a non-intentional component of consciousness was denied by many philosophers of mind after Brentano's analysis of intentionality and his pronouncement that consciousness is only 'consciousness of' (Brentano 1995). In other words conscious experience is completely described by what consciousness is experiencing – the redness of the rose, the rich bouquet of the wine and so on. There was nothing beyond content, no subject of the experience. Such a position avoids the dangerous question as to the existence of an inner experiencer, who could be regarded as acting as an inner homunculus secreted somewhere in the brain. But this leads to the well-known infinite regress of such inner homunculi, each experiencing the experiences of the preceding one (Zahavi 2005); we have discussed this already in the book, but now need to go deeper.

Western phenomenology has proposed a way of cutting off this infinite regress by stopping it before it has started: the inner homunculus is supposed to have no content, but only functions, in some as yet unknown manner, to provide the experience of ownership of that content (Zahavi 2005). This emasculated homunculus is termed the 'inner self' or the 'pre-reflective self', always being present in any conscious state. How this aborted regress does achieve its purpose of producing consciousness, as we know it from our inner experience, is presently completely unclear. For how may simple ownership alone, through the pre-reflective self, create a mechanism powerful and all embracing enough to create the overall conscious experience? There is supposedly no content in the inner self. It is bereft of all but being the owner. How can it know what it owns if it has no content? This apparent impossibility of pushing the inner self too hard to make brain-based sense of it may be why some philosophers such as (Zahavi 2005) would claim that the inner self can never be found in the brain. If it were, then its physical mechanism involved should become revealed. That appears impossible, with no such mechanism available.

We cannot thus find any approach to consciousness in the brain on a direct reading of the inner self as being content free. Any inner self that can be created by the brain must have more than the 'nothingness' of Jean-Paul Sartre, for example. One reading of (Sartre 1943) is that consciousness consists of a 'nothingness' component, which he termed 'pour-soi', and a content component he termed 'en-soi'. This certainly cuts off the infinite regress mentioned earlier, but also suffers from the criticism raised above: why does this two-component approach lead to consciousness. As I have been asked by colleagues: where is the magic that makes consciousness conscious? It does not help to try to answer this question by requiring the second component of the aborted infinite regress to have no content. Just exhausting it like a partly-used bottle of wine will not preserve the wine for a later day. The day has arrived when one is asked as to where is the magic to achieve consciousness? It just does not seem present. We must attempt to find it through a suitable brain-based model.

We have introduced in the previous chapter the attention copy signal, as the base of the CODAM model, that was suggested as being the source of the ownership signal alone in experience; only such information is supposedly held in that copy signal. However it is unclear how this signal then produces the full and rich experience of consciousness by having acted as a sentry at the gate of the buffer working memory. As has been asked already above, where is the 'magic' to breath full consciousness into these activities? We turn to that next.

We have introduced the inner self, the 'I' or pre-reflective self of Western phenomenology, as the activation on its working memory module we denoted WMcd earlier, as brought about by the attention copy signal. This can occur for endogenous attention guidance as well as in the breakthrough to awareness that can be achieved by sudden salient stimuli appearing in the sensory field. For at least in vision it is known that external stimulus activation can be processed by the frontal cortex very rapidly (in just over 100 ms) and thereby is expected to influence the parietal attention signal generator rapidly enough to function in a similar manner to the endogenous attention control signal, to bias the focal movement brought about by the attention signal generator.

It is important to consider more detailed aspects of the content of the inner or pre-reflective self, as represented by the activation of the corollary discharge module of Fig. 6.3 or 6.4. As we have developed earlier, this site is the ultimate home of the pre-reflective self and, as such therefore, the heart of consciousness (Taylor 2007 and references therein). Due to the early activity in the corollary discharge buffer of Fig. 6.3 being unconnected with lower level posterior stimulus activity, it is to be expected, as noted already, that the experience of the pre-reflective self is content free. However it may not only have ownership as content; it can play a much bigger role in the creation of consciousness and as such be related to the need for complexity in the approach of Edelman and colleagues mentioned earlier (Seth et al. 2006) and also related to the GW approach (Baars 1998) and also that of the HOT model (Rosenthal 1986).

The corollary discharge buffer is closely connected, according to the architecture of Fig. 6.3, with the error module and the target buffer module; there may be even strong interaction with the IMC for attention movement generation. Thus it is possible that other colorations of the inner self can arise besides those purely of the ownership associated with manipulation of the activity on the buffer working memory site to get the content of consciousness. Thus some knowledge will occur in the Corollary Discharge module associated with the error expected to occur in reaching the desired goal, of some aspects of the attended stimulus (from the coding on the buffer memory site), and of the attention movement signal itself.

Indeed the corollary discharge is expected to be at the centre of a complex network, comprising the totality of the inner self. There must be several buffer working memory sites (for spatial aspects of visual stimuli, for object representations, for phonological coding of language, and numerous more such short-term memory stores). Thus the cortical spread of the set of corollary discharge modules will need to be quite extensive. Moreover the set of these modules is expected to form a well-connected (possibly inhibitory) network across the cortex. This net will also be supported by its thalamic and NRT underpinning, as emphasised by earlier work of the author and many others (Taylor 2003; LaBerge 1998), to ensure good connectedness to lead to only one sense of self. Thus the corollary discharge network will spread across various areas of cortex and be thereby expected to have contacts across a considerable range of cortex. It is this feature, along with its expected central position in the brain (as the Central Representation) that will lead to the corollary discharge network being sensitive to a number of features of brain information being processed, all at the highest level. This information accessible to the corollary discharge module is therefore still only expected to be content free, so Sartre's description of it being 'Nothingness' (Sartre 1943) will still be apposite, at least from the content point of view.

The corollary discharge network is thus to be regarded, from its receipt of such high-level attention-related information, as at the centre of the attention control system. It is truly the Central Representation. It is listening-in to error correction as well as to content access for report, and may know about attention control boosting (through the error correction signal) in addition to possible switches of attention to other modalities(through distracters). Thus the inner self is expected to be at the centre of control of everything of importance ongoing in the whole brain. It is the central controller itself. But it has no content – so it could be termed the 'blind controller'.

An analogy may help here. Consider a call centre (as briefly considered earlier) where a given phone call to the centre leads to a responding phone call from the centre to a given set of phone numbers. Responders to these phone calls will then perform a variety of actions according to their function: correct incipient errors by further phone calls, repel distracters further again by extra phone calls, continue activity on the buffer site if it is there already and is desired to be continued, and so on. The call centre is functioning in the same manner as is the inner self, as a highest level controller of various attention functions.

Such a role is consistent with the hierarchical view that sensory attention is the highest control system in the brain (above motor attention, although considerably fused with it). In these terms the corollary discharge system (the corollary discharge buffer, the error correction process, the feedback to the stimulus buffer and that to the attention signal generator) is to be regarded as the top of the sensory attention hierarchy, with the module for the inner self (the corollary discharge buffer) as the centre of such information processing, with a complex range of pre-reflective awareness of various components of the message passing involved. But first among equals would be the knowledge of ownership of the consciousness of the content of the attended stimulus about to occur, with the ancillary activities being developed using this ownership signal.

Thus the 'I's eye view is that of the ongoing processing of the whole brain. It is therefore the kingpin of the brain, but is blind to detailed content. It is indeed the Central Representation of the brain.

Such an all-embracing view would seem to be contrary to that of the 'nothingness' of (Sartre 1943) or of the majority of the researchers in Western phenomenology (Zahavi 2005). However we are able to go beyond the results of the latter since we have a specific model of how consciousness could be created by CODAM (Taylor 2007), as developed in the previous chapter. The exploration of the connectivity of the corollary discharge network allows us to extract what high-level information is available to this net and so leads to the cited result. Thus the nothingness of the inner self is imbued with all that is ongoing of importance in the brain. It is content-free since it is coded at a high level whereas content is coded at a much lower level. However the attention copy signal provides the basis for the 'I' to act as the concerned watcher of all major ongoing brain processes. The 'I' has no will of its own, but is concerned with any error correction of great import as well as being cognisant with the changing contents of consciousness as they occur and over which the 'I' stands guard.

7.3 The Relation of CODAM to Other Approaches to Consciousness

We have already presented a range of neural models of consciousness in Chap. 4 (other than CODAM). We now turn to comparing CODAM with these models to see if a better fused model might then be created.

7.3.1 Consciousness and Complexity

The call centre analogy and the result that the corollary discharge network is at the highest level in the attention control network also leads to the notion that consciousness arises due to the complexity of the calls being made at any one time. Moreover they have the highest level of information, being at the top of the attention hierarchy.

into consciousness, as in the HOT approach – of CODAM is to be regarded as the detailed mechanism to achieve this consciousness-making process This detailed mechanism involves suitable speed-up and error correction (on the main working memory buffer) employing the attention copy signal on WMcd. Such essential processing by WMcd can be regarded as some form of ‘thinking’ by WMcd about WMinput, but it is more correct to consider such a mechanism as being part of the efficient control system for attention movement. WMcd activity cannot be said to have the same relation to that on WMinput as does WMinput to the lower level feature-analyzed activity in lower level cortex. This view thus does not transfer the HOT model into a control framework, but indicates how ‘thinking’ in the HOT model style may have many forms if it is to avoid the infinite regress. One of these styles is that based on the CODAM architecture.

7.4 Conclusions

In this chapter a new view has been presented of the inner self. The approach has been based on the CODAM model of the creation of consciousness presented in the previous chapter and in (Taylor 2007). This model is based on the presence and use of an attention copy signal, arising in the CODAM model as part of the attention movement control system. This signal has been proposed as the source of the experience of ownership of forthcoming content in the attention amplification process for an attended stimulus. Such ownership is essential for consciousness; without such a feature there is no-one experiencing the content associated with the attended stimulus and hence no consciousness. CODAM also extends to attention the well-established more advanced control model for motor control (more advanced, that is, than simple ballistic control), and thereby supports a partial premotor view of attention discussed in earlier chapters.

Presently the school of Western phenomenology proposes that the inner self has no content. It is problematic as to how such a system can interact with content in any useful manner. However through CODAM we can discern other forms of information of ongoing processing in the brain, all coded at a high level (so having little content). From the CODAM point of view the inner self is realized to function as the highest controller in a hierarchy of attention control, and so is to be regarded as the highest level controller in the human brain. It is the Central Representation of Chap. 2.

It is also clear from the CODAM model that information is coded at very different levels in the brain so that just using information transfer and non-linearity criteria (as do the group of Edelman, Seth et al. 2006) will not be enough to single out conscious from non-conscious neural activity. However such an approach, as well as that of the GW model of (Baars 1998) and the HOT model of (Rosenthal 1986) are all seen to be related to the overall CODAM model and provide valuable insights to help clarify the model.

References

Baars B (1998) A cognitive theory of consciousness. Cambridge University Press, Cambridge
Brentano F (1995) The psychology from an empirical standpoint. Routledge, London
LaBerge D (1998) Defining awareness by the triangular circuit of attention. Psyche 4(7). http://psyche.cs.monash.edu.au/v4/psyche-4-07-laberge.html
Rosenthal D (1986) Two concepts of consciousness. Philos Stud 49:329–359
Sartre J-P (1943) Being and nothingness. Routledge, London, 2001
Seth AK, Izhikevich E, Reeke G, Edelman GM (2006) Theories and measures of consciousness: an extended framework. Proc Natl Acad Sci USA 103(28):10799–10804
Taylor JG (2003) Paying attention to consciousness. Prog Neurobiol 71:305–335
Taylor JG (2007) CODAM: a model of attention leading to the creation of consciousness. Scholarpedia 2(11):1598
Zahavi D (2005) Subjectivity & selfhood. MIT Press, Cambridge, MA

Chapter 8
Does 'I' Really Exist?

8.1 The Problem

We are concerned here with a particular component of consciousness, its phenomenal aspect. This was already discussed in the previous two chapters (although mainly in Chap. 7), assuming it existed. The problem we now have to face is: does 'I' really exist or not?

In general the term 'consciousness' can be applied either to so-called creature consciousness (as intransitive, involving an animal who is awake, or transitive, with consciousness of such and such) or to mental state consciousness, (involving a mental state claimed to be conscious) ("Higher Order Theories of Consciousness" published in Stanford Encyclopedia of Philosophy, 2007 http://plato.stanford.edu/entries/consciousness-higher/). Here we will restrict ourselves to the detailed nature of phenomenal consciousness. In particular, the 'Problem of 'I'' of the title is that of its very existence: does there really exist an inner self, labeled by 'I'? In other words, if I am having a conscious experience, say of the delicious taste of a glass of Merlot or the odour of a rose, can I detect a further component of activity-based experience in my mind beyond that of the content of the experience I am having? Is there not only a conscious experience of that taste, smell or whatever sensation but also some knowledge that **I** am having that experience – in other words of the knowledge of the existence of my inner self as the owner of the experience? It was assumed in the previous two chapters, without argument, that such an 'I' existed. Here we have to face up to the possibility that it does not.

Such knowledge of an inner self may indeed be minimal but unless it is there, many have argued, at least from the time of Kant (if not before) that there would otherwise be no experience at all: in other words no consciousness. However there is also a flourishing philosophical community, going back to the Scottish philosopher David Hume (if not earlier) who, on the simplest reading, denied the existence of any such 'extra self'. There are also those who alternatively claim that any such experience is created in the brain solely as an illusion (Dennett 1991). There are

J.G. Taylor, *Solving the Mind-Body Problem by the CODAM Neural Model of Consciousness?*, Springer Series in Cognitive and Neural Systems 9,
DOI 10.1007/978-94-007-7645-6_8,

also many in the neuroscientific community who would agree with one or other of these 'absent 'I'' positions. Set against those who would claim they experience such an inner self or 'I', who is correct?

This is a problem that is now coming to the fore when comparisons are made between the different approaches to the nature of the mind and especially of consciousness. The whole spectrum of possibilities is now being considered, from those who state categorically that there is such an inner experience, and they report in detail on it, to those who claim they have no such inner experience and all is tranquil and silent in their mind's interior. Indeed they might even state categorically that there is no interior to their mind – all is on the surface as the content of the experience – it is all just bundles of perceptions, as Hume would have it. Thus it would appear, in this no-owner view, that in the standard five modalities there will only be brain-based activities representing to a subject their experience of the outside world but there would be no owner of those experiences. There would thus be no subject per se. The difficulty of the possible absence of 'I' is compounded by the fact that even researchers like Kriegel (2003, p. 107) state unequivocally that he is unable to offer a proof that such an inner self exists, but only achieves the statement "if there is such a thing as intransitive self-consciousness then its structure is such and such."

For those who are trying to build brain-based neural models of consciousness, and even go beyond that to create conscious machines, there is a great deal of uncertainty as to what it is they should try to include in any model they construct. Even those who wish just to give a general model of consciousness, without all of its detailed nuts and bolts exposed, will still be unsure as to what it is they are meant to model. Does it include the experience of the content as experienced by an owner or is it just the brain activity of the content?

For there to be a well-grounded science of consciousness it is important to obtain the answer to the problem of 'I'. Only then can brain imaging about consciousness, for example (in spite of its considerable progress) be given a detailed target to aim at, with the brain activity observed being realized as either only that of the content of experience or as that content plus an important component corresponding to the owner of the experience of that content. These two extremes would thereby be expected to lead to quite different resulting brain activity.

8.2 The Nature of Self or 'I'

There are many definitions of the self, especially those developed over the last few years. Thus in (Neisser 1988, p. 35) five types of self were distinguished whereas 11 years later in (Strawson 1994, p. 484) 21 concepts of the self were enumerated. All but one of this plethora of selves are not necessarily incorrect, but they are differentiated partly by the contexts in which they operate.

One aspect of the self which we are not considering here is that of the 'reflexive self', that answers to the question of 'what are my objective properties, such as:

do I have a beard, etc, etc?' This has led to brain activity studies of where such characteristics are sited in the brain. The answer is in the central components of the cerebral hemispheres and nearby areas (Feinberg and Keenan 2005), as might be expected as a site least vulnerable to brain damage. A similar set of careful discussions of the reflective self has also recently been published (Feinberg 2011); again these studies concentrate mainly on the self as possessing certain properties. It is not this directly observable aspect of the self, although it is an important component of the self, which concerns us here.

There are two basic concepts of the self that lie inside or behind such a reflexive or surface self, and that will be considered here. Both are noted as more of an 'inner' self that is behind our more directly observable self. The first concept of self is that as a narrative construction, where the self develops over the years for an individual as a way of responding as consistently as possible to the strains and stresses of everyday life. To quote (Jopling 2000, p. 83): "Selfhood is best viewed as a kind of ongoing project that serves as a response to the question of how to be". This is also well described in (Zahavi 2005, p. 105) "Who we are depends on the story we (and others) tell about ourselves". Some researchers, such as Dennett (1991), even go as far as requiring the narrative be created in language, so would be unavailable to other animals besides humans and only to children above about 2 years of age (the beginning of their language explosion).

The second approach to the self leads to the notion of the core or primitive self. This is more subtle than the narrative self and has considerable overlap with the transcendental ego of Kant, as indicated by his discussions on the topic, especially in Ch. 1 of Book II (The Transcendental Dialectic of his Critique of Pure Reason; see also the comments of Brook 2008), involving a crucial integration of self and experience. In particular the core self is to be regarded as the first person given-ness of all experiences. It accompanies these experiences by giving first person access to the experiences. Thus the primitive self is an integral part of our conscious life without which we would not be conscious of our experiences. It provides the 'what is it like to be a bat?' of Nagel, making an experience conscious as seen by the concomitant primitive self or 'I'. Without such an 'I' there can be no knower of what it is like to be, and the question cannot be answered.

There are those, such as (Dennett 1991) or (Metzinger 2003) or even earlier the well-known and rather hackneyed example of David Hume (1888), all of whom deny the existence of an inner self of the core or primitive form. Thus (Hume 1888, p. 252) "For my part, when I enter most intimately into what I call myself, I always stumble on some particular perception or other, of heat or cold, light or shade, love or hatred, pain or pleasure. I never can catch myself at any time without a perception and never can observe any thing but the perception".

Likewise in (Metzinger 2003, p. 1) he writes "no such things as selves exist in the world. Nobody was or had a self". To Metzinger any self-like entity arose from an underlying information processing model. The self itself was thereby an illusion, with any claim for its reality dashed by the fact that its character arose from it purely being a model, however one being confused with reality by a subject convinced of the existence of their 'I'.

In a similar manner Dennett denies the existence of any such inner self in his voluminous writings, including his well-known volume, majestically-entitled 'Consciousness Explained' (Dennett 1991). There he unambiguously claims we are all zombies, with no inner self whatsoever (Dennett 1991, p. 406).

It is the controversy between the 'no-inner-self' view of experience and that possessing a primitive or core self with which this chapter is concerned. It could be considered that the no-self view corresponds more closely to a 3rd person than a 1st person approach to the self. However that is untrue, since the experiences of Hume, Dennett or Metzinger are still private to themselves.

The assumption or belief that there is an inner self, an extra component of consciousness, can be traced some way back, at least as early as Kant (1956). As written in (Kriegel 2003, p. 111), Brentano (1874) developed this form of the thesis:

"A conscious perception is itself perceived; so it is perceived either by itself or by another perception; but it cannot be perceived by another perception, since this would lead to an infinite regress of perceptions; therefore it must be perceived by itself." Kriegel uses this argument to support his thesis that accompanying each perception is a parallel and inseparable component of the first perception which allows us to perceive that perception. This second perception is thus a perceptual version of the inner self.

Many philosophers have agreed on the existence of an inner self, although of a variety of characters. That the inner self is different from the initial experience is held by those such as Armstrong (1968), Carruthers (1996) and Rosenthal (1986, ibid 1993). There are also those philosophers of mind who hold the view as expressed by Brentano, that there is only one thought or experience in total, of which the inner self is a component. This group includes Brentano (1874), Husserl (1964), Sartre (1943), Sokolowski (1996), Zahavi (2005) and numerous others.

We will not attempt to follow the arguments in (Kriegel 2003) to prove that the inner self, if it exists, is only a component of the overall conscious experience, and not separate from it. Our purpose here is to attempt to address the deeper problem as to whether or not the inner self exists in the first place and try to understand its more detailed structure or composition if it does. As part of that discussion it may be necessary to call on specific components of the overall experience.

We add that the inner self we are considering is supposedly involved in making conscious the mental states of an animal brain. Thus it is to be regarded also as a component of any creature consciousness of the animal, be it transitive or intransitive. As such, therefore, any detailed discussion of any form of consciousness whatsoever is expected to have the inner self as a component. Our discussion will, however, focus more on the mental state aspect, concerned with the mechanism by which a single such state, regarded as generated by suitable neural activity in the brain, can be claimed to be a conscious state of an experiencing subject.

8.3 A New Approach to the Inner Self

So far we have looked at the inner self, or its lack, mainly from the point of view of the overall mental experience itself, be it simply of content or be it composed of a further component denoting 'I am having this experience', however faint that is. However we can also appeal to external information to see if there is any way of discovering, in an objective manner, if the inner self has any physical activity in the brain indicating or acting as a clue to its existence. Such a possibility does not seem to have been explored at all, except in recent papers of the author (Taylor 2002a, b). In this section we will attempt to determine if such hard physical data is available.

We have already reported on the disparity of views amongst philosophers of mind: some consider they experience an inner self, others dismiss it as an illusion. It would be as if, for the latter group, there was claimed by others to exist a 'Cartesian Theatre' where it would all come together in the mind at the inner self and which would be anathema to them. These doubters had themselves no such experience, so denying the inner self's existence completely. This disparity of views can be reconciled by assuming that there are different experiences for different people. Perhaps these differences are genetically laid down or correspond to character traits that have developed as they have grown up.

Such possibilities could be true. But that does not allow us to use the cases of those who claim to experience an inner self as proof that such a self can exist. For they may all be deluded in their experience and, for example, be misinterpreting typical brain-based information processing when they supposedly experience their inner self.

Hence we cannot depend on the beliefs of either the inner self believer or non-believer. Both are suspect, as arising from a possible misinterpretation of their own experience. But due to the privacy of inner experience we cannot check on the presence or not of any misinterpretation. We therefore have to look elsewhere for possible support for one or other of the position on the inner self.

What could be the character of appropriate experiments to perform on human subjects to detect an inner self? We can accept that consciousness only arises in the presence of attention, at least in many common-or-garden psychological tests. So we should follow attention tests, such as those involving speeded responses to central or nearby targets, or the attentional blink test (Vogel et al. 2005), where a rapid serial visual presentation (RSVP) stream of stimuli is presented to a subject and they have first to detect a given stimulus (first target or T1) and then detect a second one (T2) shortly afterwards. If the time between the detection of the two stimuli is about 270 ms there is a maximal difficulty of a subject (in general) from detecting T2 after a successful detection of T1. This paradigm tests attention powers maximally and has proved a valuable approach to probing attention and consciousness (especially of T2). Even so, there is no reason that we will be able to differentiate the possibility of the existence of an inner self or not by looking at brain activity. For that we need to develop some theoretical framework.

Before so doing we note that our brain-based approach is reminiscent of some ideas of Dennett in his 'heterophenomenology' (Dennett 1982). This approach is defined as corresponding to a brain-science based analysis of the beliefs of subjects whose brains are being imaged. Of course such an approach is standard in modern brain science, with subjects constantly reporting on the experiences they have in various stimulus-dependent conditions. Such reports play a basic role in the experimental paradigms now in common usage. Thus we do not dwell on the validity of such an approach, except to emphasize that it allows an experimentalist to relate the brain areas active during the information processing reported on by the subject to the actual information processing itself. In this manner considerable progress has been made in following the information flow in the brain, thereby determining various possibilities for the functions of the various brain areas whose activities are observed (not only in its spatial distribution but, as importantly, in its temporal characteristics). Such observations are, we suggest here, of crucial importance in supporting (or not) one or other of the various brain-based models of consciousness which have recently been proposed.

We should note here also that the inner self may be regarded by some as involving aspects of creature consciousness. Indeed the inner self is described as always present during creature consciousness or awareness, as noted at the end of Sect. 8.2. Of course it must be so, and hence will be present all the time a creature is aware; it is this specific feature which is regarded by the 'inner selfers' as what gives consciousness its special character, that of giving us an inner, private mental world.

8.4 Attention, Consciousness and CODAM

It could be claimed that in making the step of adding the 'owner' module in Chap. 6, so resulting in the CODAM model, we are committing the classical 'homunculus' fallacy, and hence the danger of bringing in the need for infinite regress. However the attribution of 'ownership' to such a module is minimal, and in any case cannot be said to involve consciousness of its activity. More specifically the neural activity on the Owner module of the CODAM model of Chap. 6 is that for removing distracter activity and possible amplification of the incoming target activity. We note moreover that the relevant activity on the Owner module is coded at a high level, and not connected or fed by lower level feature codes of the target or distracters present in the input stimulus. Thus this activity is expected to lead to absence of any content for the owner activity, and would lead to the characteristic 'absence of content' if a subject were able to exclude all but such owner activity from their visual processing. It is precisely that – or the associated 'stillness' – which is reported by practised meditators when they are in the state of what is called 'Pure Consciousness'.

It is important to note that the extension of the architecture of the extended ballistic control in Chap. 6 to the CODAM model is the minimal such extension. There are numerous control architectures, heavily used in control engineering applications (Smith controllers, etc, Phillips and Harbour 2000) but that of CODAM is the least possible beyond the extended ballistic control model of Chap. 5. Such an architecture also seems to be the most successful in modeling motor control in the brain (Desmurget and Grafton 2000; Wolpert and Ghararhamani 2000; Sommer and Wurtz 2002). Appealing to the pre-motor theory of attention (Rizolatti and Craighero 2010) or more generally to the efficiency of evolutionary pressures (Darwin 1859) would lead to the expectation that attention had been extended in higher animals from the architecture of the extended ballistic control model to that of CODAM.

The possible interpretation of the architecture of the CODAM model as a basis for consciousness has been explored in numerous papers of the author over the previous decade, such as (Taylor 2000, ibid 2007, ibid 2010a, ibid 2010b), and also discussed in Chap. 7.

We note that the CODAM model does not imply any self able to be probed by introspection. For the ownership activity, as noted earlier, arises only implicitly, although we must disagree with the quote from (Kriegel 2003, p. 104) when he claims the process of ownership activity is 'inattentive'. For the ownership process is part of the attention control activity, as CODAM describes more specifically, although the relevant ownership activity occurs before the awareness of the target activity arriving successfully at the working memory module. Moreover the inner self does not possess any control powers for initiating activity, only of monitoring and error correction, such as involved in rehearsal (Taylor 2002a, b, 2007; Korsten et al. 2006).

8.4.1 Detecting the Core Self

So far we have presented the CODAM architecture as being the minimal extension of the well-tried and trusted extended ballistic control architecture of attention of Chap. 5. We have also remarked how this extended CODAM architecture could be seen as somehow supporting a core or inner self, acting as an overseer of incoming activity to the 'Report' buffer of the architecture. We still have to solve the problem of obtaining evidence for the existence of the related corollary discharge signal being sent by the attention signal generator in the CODAM architecture and its possible working memory buffer. This we will do in detail in Chap. 9. The latter report module may be a luxury, but the corollary discharge signal itself certainly is not: the whole of the CODAM approach stands or falls by its existence (as would therefore any core self suggested as arising through attention).

To tie a core self to the existence of a particular corollary discharge signal, not yet observed, could be dangerous for the enterprise of discovering the core self in the brain. But yet it seems difficult to consider other identification possibilities for the core self. There is the constraint that attention is almost certainly involved in the creation of the core self in normal situations: this gives us a clue as to what we have to play with. Indeed from that point alone we should look into the possibility that the core self is created by the corollary discharge of the attention movement signal. There are other areas worthy of exploration, such as neglect and the loss of consciousness, or anosognosia (the loss of a sense of self through neglect), although we will not go into such an analysis here.

Further there are the well-attested experiences of schizophrenics, who will say 'I have lost my I' or 'I am losing my I', and are increasingly remote from the real world. In (Taylor 2010b) a detailed analysis has been given as to how CODAM helps to explain such symptoms in terms of the loss of core/inner self due to the gradual breakdown of the patient's attention control system. Without a model like CODAM, containing the possibility of coding neural activity for a core or inner self, it appears difficult to explain how such a loss could occur. The loss of a narrative self would be expected to arise in Alzheimer's disease, especially with the initial loss of hippocampus cells at the basis of long-term memory, but its loss in schizophrenia would be a mystery by such an explanation (loss of memory is not necessarily a crucial symptom of schizophrenia). We will discuss schizophrenia from the CODAM viewpoint more fully in Chap. 10.

We must also address the question: does the existence of a corollary discharge of attention actually imply the existence of a core self or could it be that we are only observing some information processing system at work, as for example, Metzinger, Dennett and Hume would have it? It is possible that a careful study of the loss of the sense of self in schizophrenia, associated with suitable brain imaging experiments, may allow us to make some progress on that. But even then the schizophrenics under study could also be mistaken about their inner experience.

It seems difficult to claim, however, that we are all, every one of us, mistaken about our inner experience. Such a level of mistakes does not change the fact that we each have that experience, however it is caused in the brain. Yet there are those who detect only bundles of perception, whilst there are others who also have an inner sense of ownership of that experience. It would therefore seem that there are still two interpretations of that inner experience – as solely a bundle of perceptions, or alternatively as having a little more, associated with our ownership sensation. The first group would expect brain architectures like that of Fig. 6.2 in Chap. 6; the second group would go for that of Fig. 6.3 or the full CODAM model of Fig. 6.4. So finally it is the extra ownership experience, and its associated brain activity, for which we must search. We must therefore ask: do human brain architectures possess (or not) a corollary discharge of attention movement signal? They are clearly differentiated by the presence of the corollary discharge.

We thus arrive at an important conclusion: there is a specific feature of brain functioning which would give rise to a sense of an inner self in the presence and use in the brain of a corollary discharge of the attention movement signal. This corollary discharge is very likely a subtle and distributed signal to search for. But it is that signal which we need in order to relate it to the creation (or loss) of a sense of 'I'. Without that signal such an aim seems far more difficult to achieve.

What evidence is there for the existence of a corollary discharge of the attention movement control signal? Let us suppose that various tests and associated brain imaging results during subject response have been collected in a particular brain imaging experiment. We would like to be able to detect in this data some specific component of brain activity and/or behaviour which accompanies the primary activity producing the experience of content (say of the second target T2 in the attentional blink), but be beyond that of normal processing under a ballistic attention control system.

One important feature we could focus on is that of the timing of the occurrence of consciousness. It is now well recognized that consciousness of a stimulus takes at least about 400–500 ms to be created in the brain post-stimulus. This creation is especially associated with what is termed the P3 peak, a positive ERP signal observed in the posterior cortices where code to be able to store the stimulus representation is being activated. Brain activity much before that time, say at 200–350 ms post-stimulus, would thereby be regarded as ancillary to the oncoming conscious experience. Of course there is important activity in posterior cortices on onset of a stimulus, where the various features of a stimulus are encoded by feature analysis in the hierarchy of visual cortices, for example. There would also be important attention feedback activity from parietal regions to amplify the relevant activity associated to an attended stimulus which is coming into consciousness by the relevant attention amplification (Gregoriou et al. 2009; Bressler et al. 2008). However these activities can be separated out from other activity, such as an attention signal to regions other than the posterior cortices (which in any case are active as analyzing the input stimulus activity, as mentioned).

Earlier feedback effects, such as of the N1 or even the C1 should not be ignored. However these occur at lower cortical level than, to us, the more important processing of the activity about to enter the buffer working memory and prior to, or at the same time, the removal of distracters. Thus we concentrate on the ERPs denoted as the N2pc, the RLIP and the SPCN (to be discussed in more detail in the following chapter), due to their activity being of greatest interest in our search for the corollary discharge of the attention control signal; these ERP signals are at a high enough cortical level to be most crucially involved in consciousness creation, but stand as possible candidates for being a corollary discharge of the attention movement control signal.

We must look, then, for any relatively early (in time) attention-based activity which we interpret as generating the experience of the inner self. This activity could allow for the creation of the experience of 'I'. The activity on which such creation would be based would be generated by the original source of the attention control signal. The main part of this signal would be sent to the back of the brain to amplify the stimulus activity and thereby bring it into consciousness (as being able to be reported to other sites in the brain). Before then a copy of that attention signal could be used for various useful purposes to make the attention amplification more efficient.

It has already been mentioned that such efficiency is well known in engineering control theory. A module called an 'observer' – a copy of the plant – has been introduced in control theory to model what is going on in a plant being controlled (Phillips and Harbour 2000). Prediction can then be made, using the observer, as to what would happen if a particular control signal were to be used on the plant. This would give a fast response able to prevent error: if the observer said 'NO, that is a dangerous control signal to send’, the control signal can then be corrected as it is performing its control function. Such error-correction would be of enormous value in the brain as it is in engineering control, so one would expect that nature had employed it in the battle for survival over the past aeons of life on earth.

At the same time such a signal could be used to prevent distracters (other stimuli) accessing the special site for report (the site of the creation of consciousness) of the stimulus representation. Such inhibition of distracters is important to prevent errors of recognition. Such errors may occur if attention is distracted, for example. Such is the cause of many car accidents. Early activity based on a copy of the attention movement control signal, which is predicting what the attended stimulus activity is going to be like, and eliminating distracter activity at an early time – as an observer – would therefore lead to a more efficient attention signal.

A copy of a control signal in general is called an ‘efference copy’ or, if the signal has been transformed into something closer to object code activity, it is termed a ‘corollary discharge’, the term we have already used. Such copy/discharge has been observed in the brain involved in motor control of eye movement (Sommer and Wurtz 2002) as well as considered as present in a number of other motor control situations (Desmurget and Grafton 2000).

We have already discussed aspects of this corollary discharge of the attention movement signal and presented the CODAM architecture that implements it in Chap. 6. The notion of a corollary discharge of the *attention* control signal is what is at issue here: attention is regarded as the main gateway to consciousness from many experiments.

We thus arrive at the question: is there any evidence, or could such be obtained, that shows a corollary discharge of attention present in the brains of subjects as they perform various attention movement tasks? Does the corollary discharge of attention exist? If so then we could suggest that it is at the basis of the creation of the experience of an inner self. It would also be helpful to be further guided by neural models of consciousness, such as they are, so as to indicate if there is any consensus on what should be observed. We will turn more directly to what we claim is strong evidence for the attention corollary discharge signal in Chap. 9, but give a more general account in the next section.

8.5 A Final Scientific Resolution

8.5.1 General Comments

Here we have discussed the nature of scientific evidence which would allow clues as to brain activity indicate the existence of an owner or inner self related to brain activity associated with consciousness. In particular we argued in earlier sections that we could not depend on the report of philosophers of mind about the existence of an inner self or owner of the content of their consciousness. This was because there were two opposing camps: those who denied the existence of any such owner, and those who strongly claimed they had experienced it. Neither of the two extreme positions was thereby proven by such reports, since the reporter could well have been mistaking their mental experience as that of an owner who does not actually exist.

Such absence of any scientific and fool-proof way through inner report of proving the existence of an inner self led us in the previous two sections to conclude that it is necessary to see if there is a suitable component, in companionship with an attention movement control signal, that could be interpreted as a signal of the ownership experience provided by any possible inner self. The nature of the search is thus for a specific signal of brain activity at specific times after a stimulus is being attended to.

We presented earlier in this chapter, and even before in Chap. 6, a basic component of brain activity suggested as helping to subserve the presence of an inner self or 'I': the corollary discharge of the attention control signal. This component was generated from a copy of the attention movement control signal. Such a brain signal has a specific signature indicating its presence: it is an extra signal from the source of the attention movement control signal. The site of the attention control movement signal itself is now well accepted as being in the superior parietal lobe and intra-parietal sulcus from numerous brain imaging experiments on attention (Corbetta et al. 2008). What is now needed is further detail of related signals which are sited in nearby cortical modules and correlated with the original attention movement control signal. There is already some initial data on this, which we will now report.

8.5.2 *The N2pc*

The N2pc derives its name (negative posterior contra-lateral activity at around the N2 signal's appearance in the ERP) from its observation by means of the attention-based early processing by attention of a visual stimulus. This occurs at 180–250 ms post-stimulus (Kiss et al. 2008) and can be observed by a more negative ERP value in the contra-lateral hemisphere in the above time period when attention is directed to a stimulus in one or other of the hemispheres. The N2pc is observed to have certain important properties, indicating it is involved in the early movement of attention to a stimulus, either by inhibiting distracters in the opposite hemisphere or amplifying target stimuli in the given hemisphere or possibly by performing both functions.

The N2pc can be observed in a more spatially refined manner by MEG (Hopf et al. 2000) where it is discovered to possess at least two components: an early one in the parietal lobe at about 180–200-ms post-stimulus and a slightly later one in the temporal lobe at about 220–240 ms post-stimulus. The first of these could correspond to an early sign of the attention control signal expected to be produced at about then, to be sent back to the posterior cortices for amplifying the target stimulus representation there. The later one has been suggested as involved in the amplification of activity entering the short term working memory, as well as inhibiting putative distracter activity there. This would therefore be exactly what is expected of the corollary discharge – related neural brain activity in the CODAM model: the first signal of the N2pc is that of the attention control signal generator, with the corollary discharge signal being seen in the second MEG activity. The timing of the second signal is after that of the first (as it should be in CODAM) and agrees with the observations of such a causal flow by fMRI in (Bressler et al. 2008) and by multiple single cell recordings in monkeys by (Gregoriou et al. 2009).

8.5.3 The RLIP Component

The RLIP component of the ERP signal was recently reported in (van der Stigchel et al. 2006). The crucial component was observed by EEG during the response by subjects in a psychological paradigm involving covert detection of a target amongst distracters. The paradigm involved a prior arrow cue indicating not only the direction of a target but also that of a distracter. As the authors stated (Sect. 3.4) "The reversed polarity supports the idea that the RLIP is a counterpart of the EDAN, not reflecting activation but inhibitory processes evoked by the distractor cue."

The RLIP signature in the ERPs was thus interpreted by the authors as arising from inhibition of the distracter position. What is important here is that this RLIP activation must have partaken of the attention control movement generation system in the superior parietal lobe, as noted in the quote above taken from the authors. For the RLIP signal requires knowledge of which of the visual stimuli presented to a subject were distracters and which was the target to be attended and responded to. The RLIP signal occurred at a still early stage in the information processing, just a little after the main attention feedback signal itself began, but slightly later than the early N2pc observed in (Hopf et al. 2000) and reported above. This timing gives the RLIP signal time to use some components of the attention movement signal generated in the superior parietal lobe and observed in (Hopf et al. 2000) at 180–200 ms.

We regard this data as further proof of the existence of the attention movement corollary discharge signal. This is seen by the RLIP activity arising at an intermediate stage in the attention processing time sequence, as just noted, and also occurring at a relatively high level in the processing (in parietal lobe). In particular the RLIP involves information from the goal, assumed to be biasing the attention control signal generator region in parietal lobe. Thus it has the general character of a corollary discharge and so the RLIP could be taken to correspond to the corollary discharge signal from 220 to 330 ms.

8.5.4 The SPCN Component

We have centred our search on the two ERPs associated with the N2pc and the RLIP. These are signals with different polarity, topography and latency. That is quite possible, corresponding to more than one component of the corollary discharge signal. Such complexity of the corollary discharge also fits the experimental results of (Robitaille and Jolicoeur 2006).

The paradigm of these latter authors involved the presentation of stimuli to both hemispheres. The subject had to detect a particular letter in one of them, although there were also distracters in that hemisphere. Besides the expected N2pc (which was independent of the number of distracters in either hemisphere) the authors observed a later negativity, which they termed the SPCN and which they interpreted as arising from the further removal of distracters, being held in the visual short term or working memory in the same hemisphere as the target. Such removal was to be compared to the removal of distracters in the opposite hemisphere to the target, which was assumed to correspond to the activity of the N2pc.

Thus in conclusion, activity associated with the N2pc involves cleaning up the non-attended side of the visual field, whilst the associated SPCN signal, occurring later, was employed to inhibit distracters in the same hemispheres. Simulation of the N2pc paradigms used the V4 <–> LIP interaction for the N2pc; it is assumed that the full CODAM model (involving STVM modules) would be needed for simulating the results of (Robitaille and Jolicoeur 2006). Such differences do indicate that complex processes, carrying the signature of a high-level goal (possibly transferred to lower cortices by a corollary discharge of the attention movement signal) are occurring just before consciousness arises of a stimulus.

Both the N2pc and the SPCN require information from a suitable goal site (prefrontally-sited) to enable the subject to remove the irrelevant distracters in the relevant paradigms. The N2pc would seem to be removing distracter activity from the hemisphere opposite to that holding activity relevant to the final goal. The SPCN, on the other hand, requires removal of distracter activity from the same brain hemisphere as contains the target stimulus. This was clear from the paradigm of (Robitaille and Jolicoeur 2006), who used stimuli on both sides of a central fixation point. In one case they employed a forward masking method, in which two similar stimuli were presented as a mask just 100 ms before the two stimuli of which one was the target. In a backward masking scenario the mask was shown just after the target stimulus and its distracter (there was also a no-mask case).

The removal of the same-sided distracter in the forward masking case is expected to be easier than the backward masking case, as was observed to be the case; the SPCN lasted longer in the latter than the former case (and both were longer than the SPCN in the no-mask situation). A similar time difference was observed in the response time. It is clear that in the forward and backward mask cases information (as a form of corollary discharge of the attention movement signal) would have been required to achieve the effects observed. This gives a clear indication of the presence of a corollary discharge of the attention in this case.

8.5.5 Initial Conclusions on the Scientific Approach

The further data needed in all these cases is that tying the suspected activity of the inner self more closely to that of the overall consciousness creation process itself. There are two possibilities as to the nature of any such ties: firstly the ties may not exist, however hard they are searched for as part of a range of attention paradigms. Secondly, on the other hand there is found to be a strong correlation and causal link between the observed corollary discharge signals and those signals proposed as the basis of consciousness as being reportable (as extended activity sited on suitable buffer sites).

In the first of these possibilities we would have to admit that we have drawn a blank in the attempt to find brain activity indicative of the experience of the inner self. Such a possibility would not be conclusive, since there may be other explanations of consciousness creation, which may also have associated with them ancillary activity indicative of the experience of an owner of the conscious activity. However all attempts may fail to discover an explanation of consciousness creation where no high level neural activity could be observed as correlated in some manner with the conscious activity. Thus we should then conclude that there is no inner self at all. The reports of the many philosophers who claim to have experienced this extra companion would have to be discounted.

On the other hand we may already have discovered relevant brain activity associated with the inner self, thereby justifying its existence, as in the N2pc, the RLIP or the SPCN ERP signals briefly described above. The reports of the many philosophers of mind who have reported they possess an inner self or ‘I’ would thereby be correct. Those who have denied its existence would have been disproved.

8.6 Bridging the Gap

There are still numerous further questions that need answers in the attempt outlined above to determine the presence or absence of an inner self along the lines of that introduced by Kant, Brentano and other thinkers, and expanded and further explored under the heading of phenomenology (Zahavi 2005; Sokolowski 2000):

1. How can we set up a more formal analysis of the possible neural mechanisms that might be involved, so as to be able to make simulations of the associated paradigms (such as those described in the previous section)?
2. Can such a formal structure, or a direct analysis, clarify the nature of the inner self over time and in particular its possible constant character (the so-called ‘constant I’)?
3. Can these features (the formal structure or a direct analysis) also provide a further understanding of the possible ‘inner black hole’ character of the inner self, as supported by the reports from those undergoing the pure conscious experience (PCE) after considerable meditation?

4. Can this formal structure also help explain in some neural detail the manner in which a variety of forms of consciousness can occur (as in dreaming, under drugs, with hallucinations, in various states of mental ill-health, such as schizophrenia, etc, etc)?

We cannot set up such a formal analysis structure here as being beyond the remit of this book, but will develop it elsewhere, as well as the beginnings of answers to these specific questions (Taylor 2011). In any case there is already a detailed system of equations for the mathematical structure of the CODAM model, already used in simulating the attentional blink, as given in the appendix to (Fragopanagos et al. 2005). That can provide a CODAM-based model approach, but may be seen to be extendable to encompass some of the other models of consciousness discussed in an earlier section.

8.7 Conclusions

The result of this chapter is that the only way to reach a final conclusion on the existence or otherwise of an inner self is to turn to investigate in considerable detail the brain activity involved in attention. In particular the correlation and causal flow of activity in the higher reaches of the brain need to be explored further. This is especially so for activity associated with certain paradigms already showing the existence of some level of what can be interpreted as corollary discharge activity of the attention movement control signal. These paradigms have shown such activity as the RLIP ERP activity in EEG by certain cueing paradigms (van der Stigchel et al. 2006), the SPCN activity observed in (Robitaille and Jolicoeur 2006), activity associated with the N2pc ERP signal, especially as observed by MEG (Hopf et al. 2000), in the SPCN ERP signals observed by (Robitaille and Jolicoeur 2006), and by the attentional blink paradigm (Sergent et al. 2005). Such brain activity, according to the conclusions we have reached in the paper, deserve more careful analysis at a more sophisticated level. At the same time such signals need probing by MEG and by fMRI, the former for more detailed timing localisation, the latter in order to consider the local sites involved in a more global network dynamics. We will develop this further in Chap. 9.

We must add to the brain imaging component the further aspect of experience reported by the subjects of any experiment being analyzed. Thus we appear to be returning to the heterophenomenology of (Dennett 1982). Yet we are not questioning the truth or otherwise of the beliefs of the subjects being analyzed, but using the reports of subjects to help understand the dynamics of brain activity involved in the information processing needed to make up the subject reports. Thus it would be of interest to explore the strength of the ownership experience in the different paradigms in the (Robitaille and Jolicoeur 2006) experiment, since that would be a test of the possibility that the length of time of the SPCN determines the strength of the ownership experience.

The above conclusion, of the need to appeal to brain activity in detail, may be regarded as a negation of philosophical discussion. However it is clear that the question of the existence of an inner self has proved elusive to philosophers over more that two millennia; the log-jam must be broken in order to make progress. Considerable advances can be claimed to have been made over the last few decades in the philosophy of mind in general and more specifically by means of brain imaging for the neural correlates of consciousness. However this crucial point on the existence or otherwise of the inner self has resisted any final conclusion being drawn. This is due to a fundamental disagreement of the interpretation of their inner experiences by philosophers of mind up to the present day. Only the experimental approach, suggested in this paper, can break that log-jam, as we have argued. The details as to how that might be achieved are adumbrated in the paper, with some support for the existence of an inner self.

Acknowledgements I would like to thank D. Rosenthal, G. Edelman and B. Baars for helpful discussions on their approaches to consciousness.

References

Armstrong DM (1968) A materialist theory of mind. Humanities Press, New York

Brentano F (1874) In: McAlister LL (ed) Psychology from empirical standpoint (trans: Rancurello AC, Terrell DB, McAlister LL). Routledge & Kegan Paul, London, 1974

Bressler S, Tang W, Sylvester CM, Shulman GL, Corbetta M (2008) Top-down control of human visual cortex by frontal and parietal cortex in anticipatory visual spatial attention. J Neurosci 28(40):10056–10061

Brook A (2008) Kant's view of the mind and consciousness of self. In: Stanford encyclopedia of philosophy. pp 1–38. http://plato.stanford.edu/entries/kant-mind/

Carruthers P (1996) Language, thought and consciousness. Cambridge University press, Cambridge

Corbetta M, Patel G, Shulman G (2008) The reorienting system of the human brain: from environment to theory of mind. Neuron 58:306–324

Darwin C (1859) On the origin of species by means of natural selection. John Murray, London

Dennett DC (1982) How to study consciousness empirically, or nothing comes to mind. Synthese 59:159–180

Dennett DC (1991) Consciousness explained. Cambridge University Press, Cambridge

Desmurget M, Grafton S (2000) Forward modelling allows feedback control for fast reaching movements. Trends Cogn Sci 4:423–431

Feinberg TE (2011) Brain and self: bridging the gap. Conscious Cogn 20(1):1–172

Feinberg TE, Keenan JP (2005) Where in the brain is the self? Conscious Cogn 14:661–678

Fragopanagos N, Kockelkoren S, Taylor JG (2005) A neurodynamic model of the attentional blink. Cogn Brain Res 24:568–586

Gregoriou GG, Gotts SJ, Zhou H, Desimone R (2009) High-frequency, long-range coupling between prefrontal and visual cortex during attention. Science 324:1207–1210

Hopf JM, Luck SL, Girelli M, Hagner T, Mangun GR, Scheich H, Heinze HJ (2000) Neural sources of focused attention in visual search. Cereb Cortex 10:1233–1241

Hume D (1888) A treatise on human nature. Clarendon Press, Oxford

Husserl E (1964) Phenomenology of internal time consciousness (trans: Churchill JS; ed. Heidegger M). Indiana University Press, Bloomington

Jopling DA (2000) Self-knowledge and the self. Routledge, London
Kant I (1956) Kritik der reinen Wernunft. Felix Meiner, Hamburg
Kiss M, Van Velzen J, Eimer M (2008) The N2pc component and its links to attention shifts and spatially selective visual processing. Psychophysiology 45:240–249
Korsten N, Fragopanagos N, Hartley M, Taylor N, Taylor JG (2006) Attention as a controller. Neural Netw 19(9):1408–1421
Kriegel U (2003) Consciousness as intransitive self-consciousness: two views and an argument. Can J Philos 33(1):103–132
Metzinger T (2003) Being no-one. MIT Press, Cambridge, MA
Neisser U (1988) Five kinds of self-knowledge. Philos Psychol 1(1):35–59
Phillips C, Harbour P (2000) Feedback control systems. Prentice Hall, Upper Saddle River
Rizolatti G, Craighero L (2010) The premotor theory of attention. Scholarpedia 5(1):6311
Robitaille N, Jolicoeur P (2006) Fundamental properties of the N2pc as an index of spatial attention: effects of masking. Can J Exp Psychol 60:79–89
Rosenthal DM (1986) Two concepts of consciousness. Philos Stud 49:329–359
Rosenthal DM (1993) Thinking that one thinks. In: Humphreys G, Davies M (eds) Consciousness. Basil Blackwell, Oxford
Sartre JP (1943) Being and nothingness. Routledge, London, 2001
Sergent C, Baillet S, Dehaene S (2005) Timing of the brain events underlying access to consciousness during the attentional blink. Nat Neurosci 8(10):1391–1400
Sokolowski R (1996) Husserlian meditations. Northwestern University Press, Evanston
Sokolowski R (2000) Introduction to phenomenology. Cambridge University Press, Cambridge
Sommer MA, Wurtz RH (2002) A pathway in primate brain for internal monitoring of movements. Science 296(5572):1480–1482
Strawson G (1994) Mental reality. MIT Press, Cambridge
Taylor JG (2000) A control model for attention and consciousness. Soc Neurosci Abstr 26:2231#839.3
Taylor JG (2002a) Paying attention to consciousness. Trends Cogn Sci 6(5):206–210
Taylor JG (2002b) From matter to mind. J Conscious Stud 6:3–22
Taylor JG (2007) CODAM: a model of attention leading to the creation of consciousness. Scholarpedia 2(11):1598
Taylor JG (2010a) The I's eye view of its consciousness. J Conscious Stud 17(1/2):95–117
Taylor JG (2010b) A neural model of the loss of self in schizophrenia. Schizophr Bull (on-line: 23 April 2010)
Taylor JG (2011) A neural model of loss of self in schizophrenia. Schizophr Bull 37(6):1229–1247
Van der Stigchel S, Heslenfeld DJ, Theeuwes J (2006) An ERP study of preparatory and inhibitory mechanisms in a cued saccade task. Brain Res 1105(1):32–45
Vogel EK, McCollough AW, Machizawa MG (2005) Neural measures reveal individual differences in controlling access to working memory. Nature 438(24):500–503
Wolpert DM, Ghahramani Z (2000) Computational principles of movement neuroscience. Nat Neurosci 3:1212–1217
Zahavi D (2005) Subjectivity & selfhood. MIT Press, Cambridge, MA

Chapter 9
Does the Corollary Discharger of Attention Exist?

9.1 Introduction

The CODAM model was introduced as a model of conscious experience in 1999/2000 (Taylor 1999; ibid 2000) and developed subsequently over that decade in a series of papers (see references and those in Taylor 2007) with applications to various well-studied experimental paradigms on attention (Taylor and Rogers 2002; Taylor and Fragopanagos 2003; Fragopanagos et al. 2005; Korsten et al. 2006; Fragopanagos and Taylor 2007); it was also applied to explain a range of meditation and related experiences (Taylor 2002a, b). Most recently it has been developed in more detail in terms of underpinning neuro-modulators (in particular dopamine and acetylcholine) for explaining the experiences of schizophrenics across the four main symptoms of prodromal, positive, negative and disordered (Taylor 2011), as we will discuss in Chap. 10. We have already discussed aspects of CODAM, from its basis in the nature of attention control (in Chaps. 5 and 6) to its ability to provide an experience of 'ownership' to the subject (in Chap. 7), although the question of the evidence for that ownership was questioned (in Chap. 8). However the basic problem of CODAM over the last decade has been lack of clear experimental evidence for the existence of the basic corollary discharge of attention movement on which the CODAM model was founded. Some initial evidence was claimed from EEG results with the attentional blink (Sergent et al. 2005), but that was only the first sighting and needs further data to be able to support the thesis that the corollary discharge of attention movement does exist. The further details of its dynamics is only then to be determined in detail, possibly in combination with an underlying more detailed model of such a corollary discharge as can be supported by more detailed investigations of attention.

It is the purpose of this chapter to collect and discuss experimental evidence for the existence of the corollary discharge of attention movement signal. In so doing we hope thereby to be able to give more structure to the ongoing dynamics of visual attention in the intermediate temporal stage between the initial attention-based input stimulus-based brain activity, as very likely coded by the N2pc at about 180–300 ms

J.G. Taylor, *Solving the Mind-Body Problem by the CODAM Neural Model of Consciousness?*, Springer Series in Cognitive and Neural Systems 9,
DOI 10.1007/978-94-007-7645-6_9, © Springer Science+Business Media Dordrecht 2013

post-stimulus, or even earlier by attention amplification effects through the P1/N1 complex at about 100–200 ms, and the final stage associated with access to visual short term memory (VSTM) for report at about 500–1,000 ms post-stimulus (the exact time depending on the experimental paradigm). We will review in particular data arising from a range of MEG and EEG experiments that enable attention dynamics at such intermediate times, between stimulus input and report, to be observed using high-density electrode sampling or by use of MEG sensors (Hopf et al. 2000, 2002a, b, 2004; Robitaille and Jolicoeur 2006; Robitaille et al. 2009). In addition, the data of (Sergent et al. 2005) will be re-considered. At the same time other relevant data will be more briefly mentioned. The data will also be considered from the point of view of our putative CODAM model, as referenced above. We already gave an introductory discussion of these various specific ERP signals. Here we will attempt a much more detailed and complete analysis of evidence for the existence of the corollary discharge.

We have already considered the control nature of attention in Chap. 5 and the extension of the extended ballistic control model of attention movement, arrived at from that discussion, to the CODAM model in Chap. 6. The crucial extension here was carried by addition to the former model of a corollary discharge signal generated by the attention movement generator. Thence the full title of the new model: the COrollary Discharge of Attention Movement (CODAM). However we still do not have a full armory of experimental data under our belts to give some justification for the existence of the new CODAM model, so by the existence of the corollary discharge signal. We consider that next.

9.2 The Corollary Discharge Data

9.2.1 Criteria for a Corollary Discharge

Various properties of a corollary discharge signal must be specified before we can try to discover if such a signal has already been observed in experimental data. In other words we need roughly to know what we are looking for before we set out on our quest for a corollary discharge of attention movement in the brain. Especially we should note that the search has not been carried out before, to our knowledge, for attention movement control: no attempt by others has been made to understand attention movement using the more sophisticated ideas of modern control theory. However we will take guidance from the successful application of these ideas in motor control (Wolpert and Ghahramani 2000; Desmurget and Grafton 2000). We can also expect the existence of such a corollary discharge for attention as supported by the pre-motor theory of attention (Rizolatti and Craighero 2010). Such an approach considers attention movement as possessing considerable similarity to that of motor action, with some considerable overlap being presumed between the two sets of functional areas involved in attention and motor control.

We start with the four criteria successfully applied to the search for the corollary discharge of a motor control signal, in this case for eye movement control (Wurtz and Sommer 2004). The four criteria they used were:

C1 The signals originate in a motor area;
C2 The signals precede and spatially represent the movement;
C3 Eliminating the signals does not impair movements in tasks not requiring corollary discharge
C4 Eliminating the signals *does* impair movements in tasks requiring corollary discharge.

These criteria were applied successfully (Wurtz and Sommer 2004 and earlier papers) to experimental results on the proposed eye movement control corollary discharge signal, from the superior colliculus to the thalamus and thence to the frontal eye fields (FEF). The four criteria C1–C4 were shown to be satisfied by the observed signal in monkeys, as well as when these signals were prevented from reaching the FEF. More detail is given in the reference (Wurtz and Sommer 2004), to which the interested reader is directed.

In order to apply these criteria to the case of the search for a putative corollary discharge of the attention movement control signal, we must modify the criteria C1–C4 as follows:

A1 The signals originate in an attention control area;
A2 The signals precede and spatially represent the actual attention movement;
A3 Eliminating the signals do not impair attention movement in tasks not requiring corollary discharge;
A4 Eliminating the signals do impair attention movement in tasks requiring such a corollary discharge.

The criteria A1 and A2 can be applied directly to any observed attention movement control process. A3 and A4 require considerable care in that the elimination process itself may need to be done on humans (without damage to their brains). This can be achieved in one way, for example, by means of TMS, as has already been achieved in a number of relevant experiments. An alternative source of data is from patients with brain defects, especially in stroke: such patients are well-known to have defects in the attention system, as many researchers have argued.

We must add that there is still uncertainty about the nature of the proposed attention copy signal. Is there just one such signal to be searched for, or are there different signals being involved in different functions associated with attention (but not with direct modification of lower level sensory processing)? It is feasible that there may be a number of different functions carried out by such a copy signal: amplification of the target representation as by refreshing its working memory representation, inhibition of possible distracters entering the working memory, activating reward values for the target, being used for error correction given an incorrect target. There may also be other uses. This is consistent with the increasing set of components of higher-order information processing activity and their functionality now being uncovered in the brain.

9.2.2 Looking for the Relevant Signal(s)

We must now turn to the set of signals in the brain we may use to provide evidence for the existence of the corollary discharge of attention signal. To do that we have first to search for the attention movement control signal itself and determine its timing. That signal would be one sent to the lower posterior sensory cortices so as to amplify the neural activity representing the relevant target stimulus, with associated inhibition of any distracter, especially if they are near to the target. That will provide us with a time for any possible beginning of a corollary discharge signal.

We have already discussed some of the many fMRI signals analyzed in a range of subtle paradigms, with the control circuits of attention as developed in Chap. 5, thereby summarizing a large amount of data on attention in the brain. However, blood flow signals do not have enough temporal sensitivity to be able to single out those signals which arise on the order of tens of milliseconds apart, as is needed to test for the expected dynamical flow arising from such a model as that of CODAM. Nor are fMRI signals able to probe the several hundred millisecond delay between stimulus onset and expected awareness, due to the coarser temporal sensitivity of the dynamics of fMRI signals. This problem is even more acute for PET observations.

We are thus left with the much greater temporal sensitivity of EEG and MEG measurements. However these have an alternative problem of lack of spatial sensitivity, due to problems of possible ambiguities arising in the solution of the inverse problem (calculating the underlying current sources in the brain that cause the electric or magnetic fields on the scalp in the first place). Yet these electric or magnetic field measurement techniques, in conjunction with spatially accurate blood flow methods (PET & fMRI), can help to probe the relevant brain signals that could signal first of all the presence of an attention feedback signal and then secondly a higher-level corollary discharge or efference copy signal of that basic attention movement signal, all with the needed temporal and spatial accuracy.

We meet a further problem in such an approach: there seems to be an apparent contradiction between EEG and MEG signals as to when exactly the focus of attention can be observed to move. For example there is the set of components observed under certain paradigms described in (Kiss et al. 2008), with acronyms such as EDAN, ADAN, LDAP, RLIP, or those ERP signals detected by other paradigms, such as the P(D) and N(T), in (Hickey et al. 2009) and suggested to be components of the N2pc, or the later component denoted Ptc observed in the paradigms of (Hillmire et al. 2010), as well as the even later SPCN of (Robitaille and Jolicoeur 2006, and also observed by other research groups). Which of these signals (possibly more than one) could represent an attention movement signal? In particular EEG seems to indicate an EDAN-based attention movement control signal (Kiss et al. 2008) as compared to an MEG-based N2pc signal with that functionality (Hopf et al. 2000).

From the above plethora of ERP and MEG signals, we cannot conclude there is one and only one such corollary discharge signal to search for, or to be expected. But certainly we have to look outside primary sensory regions and after the initial attention movement control signal (whenever that is) has been sent to sensory cortices, for any signal that can be suggested as a corollary discharge of this primary attention movement control signal.

The results of (Kiss et al. 2008) would seem to indicate that only the ADAN and LDAP are good candidates for the initial attention movement control signal. These are the only signals they observed in their paradigm (to be discussed shortly) in which either an informative or a non-informative cue appeared before a target. In the former case a clear ADAN signal was seen after the cue, whilst for the latter this had vanished. Moreover when a target appeared a very clear N2pc was observed for both the informative and non-informative cue cases. This seemed to indicate, as they claimed (Kiss et al. 2008, p. 247) that attention had already moved to the necessary hemisphere in both the cued cases and before the N2pc, concluding that "... the N2pc is instead linked to processes that occur after such shifts have been completed." However (Eimer, private communication) has noted that "we are still looking for this elusive signal (of the movement of attention)" and more work needs to be done to pin it down.

This difficulty arose because the relevant LDAP posterior cortical ERP signal, in the Kiss et al case (Kiss et al. 2008, p. 244) seemed to rule out the N2pc (being the same in the informative and the uninformative cue cases) but furthermore the LDAP "was absent with uninformative cues but only revealed a trend towards the presence of the LDAP in the informative cue condition that fell short of statistical significance..." There are numerous sightings of the LDAP by other research teams, but none (known to the author) in association with the crucial informative/ uninformative cue comparison.

9.2.3 *Experimental Support for the Existence of an Attention Movement Signal from MEG Data*

We hypothesise the time line of attention movement control activity, from the extended attention control model of Figs. 6.3 and 6.4 of the CODAM architecture, as follows:

Proposed Time Line

$$0\ (\text{stimulus onset}) \rightarrow 50-200\ \text{ms}\ \Big(\text{feature analysis, possibly attention-modulated by a previous attention control signal, and further modulated by local low-level saliency processing}\Big) \rightarrow 180\text{–?ms}\ (\text{attention control feedback signal generation from parietal to lower level cortical sites}) \rightarrow 220-350\ \text{ms}\ (\text{corollary discharge, before report occurs, with some removal of distracters}) \rightarrow 300-1{,}000\ \text{ms}\ (\text{attended stimulus access to its short-term or working memory, to be available for report, and involving further possible corollary discharge activity enabling more complex distracter removal than during the earlier phase}\Big) \quad (9.1)$$

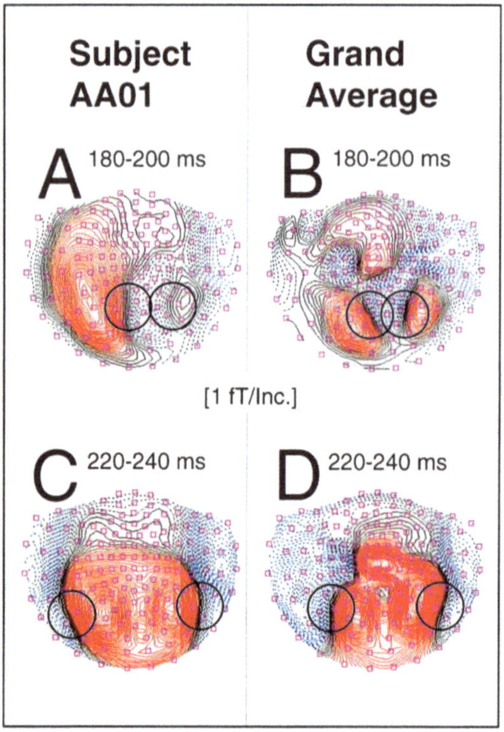

Fig. 9.1 The two components of the N2pc signal

The times inserted into the time-line (9.1) are those taken from a variety of measurements of brain activity, but can be varied over a certain range according to the paradigm. In particular they have been taken from the appropriate data we will be using to explore the possibility associated with the proposed corollary discharge. We have not been able to specify a final time by which the attention copy should be created, but assume it is before report (and hence awareness of content) occurs. This is supported by the assumed processing of the corollary discharge signal, acting as an inhibitor of all distracters in the working memory before the possibility of consciousness of the attended target activity.

There are several sets of relevant data known to us to explore the time line (9.1), and in particular the signal for the movement of attention. The first of these was obtained a decade ago (Hopf et al. 2000) and is shown in Fig. 9.1.

Magnetic field distributions over two different time ranges from the single-subject multi-session data (A and C) and the grand average data (B and D) measured from LVF-minus-RVF target difference waves (where LVF and RVF denote left and right visual field respectively). The black circles indicate transition points between locations where the magnetic field leaves (in red) and enters (in blue) the skull, which are usually associated with an underlying dipole. The paradigm consisted of two sets of 12 short horizontal and vertical lines, one set on each side of a central fixation cross. Subjects were asked to fixate on the central cross and attend to either a red or a green bar (the other 22 bars being blue distracter bars, and the red and green bars being in opposite hemispheres) and indicate whether the attended bar was horizontal or vertical in each stimulus array (with the colour to be attended being specified at the beginning of each block of 200 individual trials, with 12 trial blocks in all). The stimuli for a given trial were on for 750 ms, and there was a variable-duration gap of 600–900 ms between each stimulus presentation for response. Stimulus confounds were avoided by using red and green bars in each of the stimulus trials. A & B: The early component at 180–200 ms; C & D: The late component at 220–240 ms (adapted from (Hopf et al. 2000) with permission)

The data for Fig. 9.1 was taken (Hopf et al. 2000) by MEG for subjects performing a target search task that produced a difference signal between the two hemispheres. In particular there was a difference in the activity contralateral to the target as compared to that in the same hemisphere.

The paradigm consisted of two sets of 12 short horizontal and vertical lines, one set on each side of a central fixation cross. Subjects were asked to fixate on the central cross and attend to either a red or a green bar (the other 22 bars being blue distracter bars, and the red and green bars being in opposite hemispheres) and indicate whether the attended bar was horizontal or vertical in each stimulus array (with the colour to be attended being specified at the beginning of each block of 200 individual trials, with 12 trial blocks in all). The stimuli for a given trial were on for 750 ms, and there was a variable-duration gap of 600–900 ms between each stimulus presentation for response. Stimulus confounds were avoided by using red and green bars in each of the stimulus trials.

As shown in Fig. 9.1, the resulting difference signal between the two hemispheres was observed to have two components: one an early one at 180–200 ms in SPL and the second at 220–240 ms, observed in the temporal lobe. We take the first of these signals to show the initial attention movement control signal being generated in the SPL/IPS complex (Corbetta et al. 2008; Bressler et al. 2008). This timing data was used in the time line (9.1). The later signal at 220–240 ms may represent the arrival at a temporal lobe working memory site, which we propose to be for a corollary discharge created from the earlier SPL signal (as corresponds to the architecture of Figs. 6.3 or 6.4.)

One supposed problem with this data is that the early MEG signal has not been observed in later MEG experiments performed by Hopf and his colleagues (Hopf et al. 2002a; ibid 2004). However this discrepancy has been handled by the Hopf group by noting that the experiments reported in the papers of 2002a and 2004 very likely had 'silent' MEG signals compared to those arising in the 2000 paper. As noted in the 2002a paper:

> ...the pattern of results across experiments is consistent with the proposal that the parietal subcomponent reflects the mechanism that initiates shifts of attention, whereas the occipito-temporal subcomponent reflects the mechanisms that implement the selection of relevant information once attention has been shifted. (Hopf et al. 2002a, p 27).

This view was claimed in that paper to be further supported by the experimental results reported in (Hopf et al. 2002b). This is completely consistent with our earlier suggestion that the attention movement signal is generated as the early component of the N2pc in the parietal lobe, and used to amplify target related lower level activity so as to access the relevant short term visual memory (or visual working memory). This localization in the SPL/IPS is also supported by the results of (Corbetta et al. 2008; Bressler et al. 2008). There will be further support for this overall mechanism below, and especially the working memory component of it, when we turn to the SPCN results of (Robitaille and Jolicoeur 2006).

We still need somehow to reconcile the (Hopf et al. 2000) claim that the early component of the N2pc is that of the signal of attention movement with the contradictory data of (Kiss et al. 2008); this latter was claimed to show that the N2pc occurs after attention movement has occurred. The paradigm of (Kiss et al. 2008) used a circular arrangement of 12 small squares, ordered as if around the circumference of a clock, with a single square being oriented as a diamond, with one side of the diamond cut out. The earlier cue consisted of an arrow, pointing either to the 2, 3 or 4 positions on the clock or to those at 8, 9, 10 on the other side. In the uninformative cue case there were two such arrows, pointing in opposite directions (so indeed being uninformative).

One difficulty of this arrangement of cue and distracters is that the circle only subtended 4.5 degrees of angle to the subject from the central fixation cross, so could possibly have been accommodated by a suitably broad attention focus across the whole display. Thus in the uninformative cue case there may be a movement of attention producing an N2pc, which might even be similar to that in the informative cue case (as observed), so explaining the identity of the N2pc in the two cases (informative and uninformative cue cases).

Another avenue from which to attack this problem of discovering the brain signal for attention movement is by the use of trans-cranial magnetic stimulation (TMS) to slow down the attention movement process (Schenkluhn et al. 2008). The paradigm involved applying TMS to three different regions of the parietal lobes (supramarginal gyrus, posterior and anterior intraparietal sulcus) to determine which of these three areas was sensitive to the features of colour or spatial nature (or a neutral uninformative cue) if TMS was applied to them. An array of partial circles, with a small segment being taken from each of them (a Landolt array), with one circle of the set being complete and coloured, was presented to the subject and their task was to detect where that single complete circle was sited. After cue onset for each trial the TMS pulse was applied from 100 to 500 ms post-cue. Such timing would have been before the LDAP signal was created, according to the timings observed in (Kiss et al. 2008), but it covers the times of the N2pc signal observed by (Hopf et al. 2000). The results indicated suitable slowing, so dependence, only on the supramarginal gyrus. The timing of this effect was therefore compatible with the attention movement control signal (if that was being distorted) occurring as from the N2pc or before it.

We summarize the points pro and con for the signals (ADAN/LDAP pair versus the parietal component of the MEG N2pc signal) to represent the attention movement signal in Table 9.1 below. We have mentioned the ADAN signal briefly

Table 9.1 Which is the attention movement control signal?

Signal considered	Pro	Con
N2pc (Hopf et al. 2000)	1. Clearly observed, but absent in later paradigms (Hopf et al. 2002a, b, 2004). This discrepancy was explained in those papers 2. Suitably early (180–200 ms) and sited in the parietal lobes 3. Experimentally observed vulnerable to TMS applied to SPL (Schenkluhn et al. 2008)	1. Early parietal activation suspected by some other researchers (claimed to be due to possibly poor source localisation modelling) 2. Present in both the informative and uninformative cases as reported in (Kiss et al. 2008)
ADAN (frontal)/LDAP (posterior) Cortical ERP Signals (Kiss et al. 2008)	1. ADAN observed as significant in the informative cue case, and not in the uninformative case	1. LDAP not observed as significant in either the informative or uninformative cue cases 2. Only hinted at after a double subtraction

as a frontal precursor of the posterior LDAP signal; as such, ADAN may feed into LDAP, although no causal flow from the former to the latter has been reported.

Pros and Cons for the possibility of the ADAN/LDAP pair or the N2pc signal being the attention movement control signal, as required exists in any control model of attention.

We conclude from Table 9.1 that there is more evidence for the N2pc to be involved with the attention movement control signal than for the LDAP signal. That the former is not sensitive to the information carried by any pre-cue, as reported in (Kiss et al. 2008), may be explained by the uninformative cue acting to alert the attention focus to the whole (known to be relatively small) central target, which can then be centred on the target side when the target plus distracters appear.

In any case we can assume, consistent with all the above data, that the attention movement signal was generated at or before the N2pc signal at around 200 ms post-stimulus. Such timing has been proposed by the majority of investigators of the N2pc, including Kiss et al. in their 2008 paper. Thus we will search for the corollary discharge signal as created during or after the N2pc.

9.2.4 The SPCN Component

Recent paradigms have exposed a further important component of the ERP sequence: that of the SPCN (the sustained posterior contralateral negativity) mentioned earlier, and detected in ERP investigations of the dependence of the N2pc on masking: both forward and backward masking has been employed to study the effect on the N2pc (Robitaille and Jolicoeur 2006), as well as in earlier investigations of the ERP component related to visual short term memory (VSTM) (Vogel and Machizawa 2004). These earlier references give important features of the SPCN activity, especially in relation to the capacity of VSTM and of a subject's ability to prevent distracter information from entering and degrading short-term memory. However a more refined experimental paradigm to probe VSTM was used in (Robitaille and Jolicoeur 2006), to which we now turn.

The paradigm employed in (Robitaille and Jolicoeur 2006) used a pair of coloured letters or digits, each presented for 100 ms, one on either side of fixation (one digit and one letter were used at a time on either side of the fixation point). The target character for detection had a specific colour, with one of the two characters presented having this colour, the other character having the other colour (pink and green were the two colours employed). After an exposure of 100 ms, a further similar exposure of two (similarly coloured) new characters was presented for the same period, to act either as a backward mask or alternatively as a target, with the earlier pair of letters then functioning as a forward mask (after instruction to the subject). In the no-mask case only one pair of letters was presented, with a blank screen for the second stimulus. The colour for a character category (letter or digit) was held constant for a given subject over the entire testing session.

The figure shows the N2pc (at 180–281 ms) and the SPCN (at 301–900 ms) as observed in (Robitaille and Jolicoeur 2006) (Fig. 9.3 of that paper, permissions to be obtained). The paradigm employed in (Robitaille and Jolicoeur 2006) used a pair of coloured letters or digits, each presented for 100 ms, one on either side of fixation (one digit and one letter were used at a time on either side of the fixation point). The target character for detection had a specific colour, with one of the two characters presented having this colour, the other character having the other colour (pink and green were the two colours employed). After an exposure of 100 ms, a further similar exposure of two similarly coloured new characters was presented for the same period, to act either as a backward mask or alternatively as a target, with the earlier pair of letters then functioning as a forward mask (after instruction to the subject). In the no-mask case only one pair of letters was presented, with a blank screen for the second stimulus. The colour for a character category (letter or digit) was held constant for a given subject over the entire testing session.

The results of the experiment of (Robitaille and Jolicoeur 2006, which we henceforth denote R & J,) are given in Fig. 9.2, which shows there is an absence of any change of the N2pc caused by masking. Thus we can conclude that the N2pc (at least in the paradigm of R & J) is involved in focussing attention onto the relevant hemisphere in space for further processing to occur. In the masking paradigm used in R & J there was also the need to inhibit the distracter either coming just before (for forward masking) or just after (for backward masking) the target itself in the same position in space.

The important further component of processing, the SPCN, was observed by R & J in the ERP signal over the period of 300–1,000 ms after stimulus onset, as shown in Fig. 9.2. The SPCN was longest for backward masking (from 300 to 900 ms post-stimulus), shorter for forward masking (lasting for 300–700 ms) and shortest for no masking (300–500 ms). It was proposed by the authors that the SPCN reflected the presence of processing of the target and its mask (in the same hemisphere) inside the Visual Short Term Memory (VSTM), this processing being absent for the

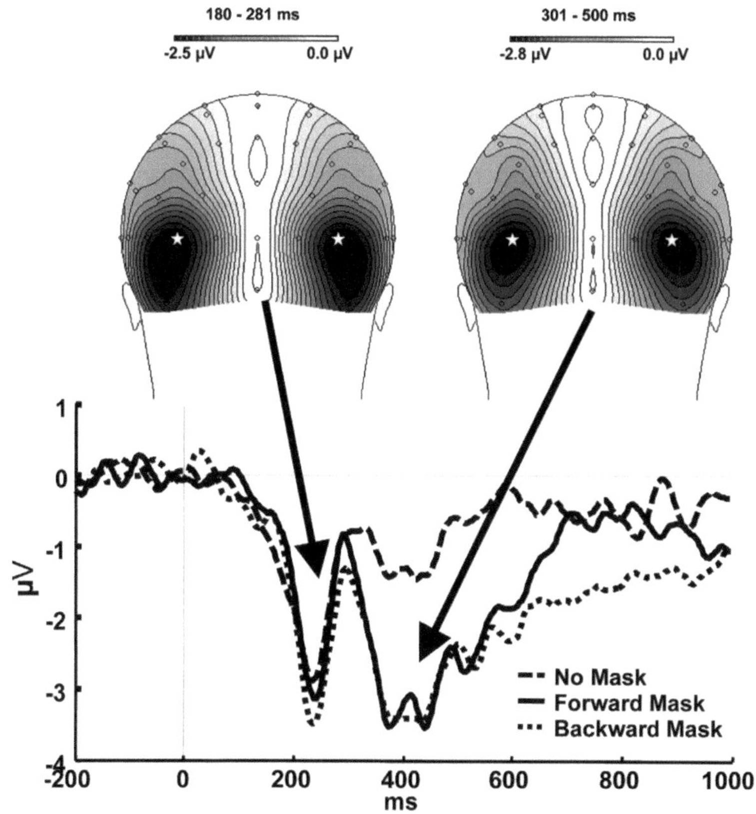

Fig. 9.2 The N2pc and SPCN as detected in (Robitaille and Jolicoeur 2006)

no-mask events. The dependence of the length of this processing indicated that such processing arose from removal of distracter stimuli: this removal was conjectured in R & J as more difficult for the backward masking case, less difficult for the forward mask and least difficult of all in the no mask situation.

The negative value of the ERP in the SPCN period corresponds, as is usually argued, to indicate removal of distracters, as does the N2pc at the earlier stage. Such inhibitory character of the N2pc has been proposed by numerous investigators (Kiss et al. 2008; Luck and Hillyard 1994a, b). The distracters during the SPCN have penetrated the visual STM, as shown by continued activity of the SPCN, especially for backward masking. The length of this SPCN negativity, as seen in Fig. 9.2, was observed to be closely correlated to the reaction times under the three conditions, these having mean values 601, 718 and 1,000 ms in the no-mask, forward mask and backward mask cases respectively (R & J).

To achieve this differential timing effect, a corollary discharge of the attention signal, carrying attention goal information, must have been sent to the working memory module to provide suitable goal information in the VSTM. Such goal information is that of removing the distracter activity in the VSTM in preference to that of the given target. The time during which the SPCN is acting appears appropriate for such inhibitory processing in the working memory module (lasting for several hundred mille-seconds), with identification of this working memory site with the Visual Short Term Memory site considered in R & J.

We support this identification of the SPCN with a component of the corollary discharge by considering relevant details of the results presented in R & J: the SPCN carries a signature of the goal of the paradigm, indicated by differences between the various temporal durations of the SPCN as correlated with the difficulty of the masking conditions, and as carefully discussed in R & J. The SPCN involves goal-biased information in the parieto-occipital regions, from SPCN activity being detected by MEG there (Robitaille et al. 2009). Such information would arise from the intra-parietal sulcus/superior parietal lobe (IPS/SPL) source of the attention control signal (Hopf et al. 2000) or directly from the goal module in prefrontal cortex, in the prefrontal cortex/frontal eye fields (PFC/FEF) (Corbetta et al. 2008; Sridharan et al. 2007; Bressler et al. 2008; Gregoriou et al. 2009). In either case we conclude that:

The SPCN signal of R & J carries appropriate corollary discharge activity of attention movement to enable removal of the distracter, in either forward or backward masking conditions.

It is possible to check the above by determining the correlation of the SPCN with the SPL/IPS and PFC/FEF activity during this distracter-removal processing period. It would also be important to use Granger causality to show the causal flow of activity from the SPL/IPS or PFC/FEF sites so as to demonstrate that the SPCN, as a corollary discharge, is definitely arising in a causal manner from these latter sites; such data is not presently available.

9.2.5 Explanations and Architectures

The neural dynamics leading to the SPCN component of the ERP curves of Fig. 9.1 could be as follows:

(a) The observed ERPs are different for the three different masks (forward, backward and no mask), with persistence times of the SPCN in the three cases given earlier. This is what has to be explained by some dynamical process;
(b) As is clear from the temporal durations of the SPCN under the three conditions, the three masked cases give successively longer durations for the associated SPCN. In the no-mask condition, there are no distracters (except from the opposite hemisphere), so the activity is expected to be shorter than the forward or backward cases, as observed in (a). In the masking cases, the forward mask SPCN lasts a shorter time than the backward mask, suggested in R & J as arising from the removal of the distracter from the forward mask task being simpler than that for the backward mask case;

(c) In the no-mask case the short (about 200 ms) but non-zero duration of the SPCN could arise from a minimal corollary discharge to remove the remainder of any non-target from the opposite hemisphere (so completing the inhibitory processing detected in the N2pc, and as suggested in R & J);
(d) For both the forward and backward mask cases, the two stimulus activities (either digits or letters of the right colour) are both under the control of attention. The earlier or later of these signals has to be deleted by use of part of the corollary discharge signal, for the forward or backward mask cases respectively;
(e) Deletion of earlier activity, in the forward mask case, can be achieved by its removal from the VSTM by a competitive (inhibitory) process there. This should cause removal of activity on VSTM other than the most active, assumed to be the target (as the latest input) in the forward mask case. Such a competition could be triggered by the corollary discharge, say by enhancement of lateral inhibitory connectivity on the VSTM;
(f) Removal of the later input to the VSTM in the backward mask case must use a more complex mechanism, such as first deleting the earlier neuron's activity by means of competitive activity (as in the forward mask case) on a copy of the VSTM module, and then sending an inhibitory signal back from that copy module to the earlier VSTM module, to remove the neural activity associated to the later (larger) stimulus. The resulting activity on the initial VSTM module will then consist solely of the earlier (and weaker) target representation.

These are feasible neural mechanisms for explaining the important data of R & J; they can be checked by suitable further experimental data, especially in the two-stage process in the backward mask case (f).

In summary, the details of the SPCN activity, before report is able to be made, are only explicable in terms involving information that could only be contained in a corollary discharge of the attention movement control signal. For such a signal must possess information about the goal being attempted, in this case detecting and removing a particular target in the masking paradigm of (R & J). It is such employment of a corollary discharge of attention which supports the CODAM model. Further experimental study of the SPCN signal is essential to enhance this understanding.

9.2.6 The Attentional Blink

There is further ERP data supporting the existence of the corollary discharge signal, as arising from the experimental results of (Sergent et al. 2005) using the attentional blink paradigm. This is especially noted in the figure 8 of their paper, repeated in our Fig. 9.3 (although with additional annotation).

Data from (Sergent et al. 2005) on the attentional blink. The details in the figure are discussed below in the text. The attentional blink occurs when a series of putative targets, such as letters, is briefly shone on a screen before a subject at a rate of about

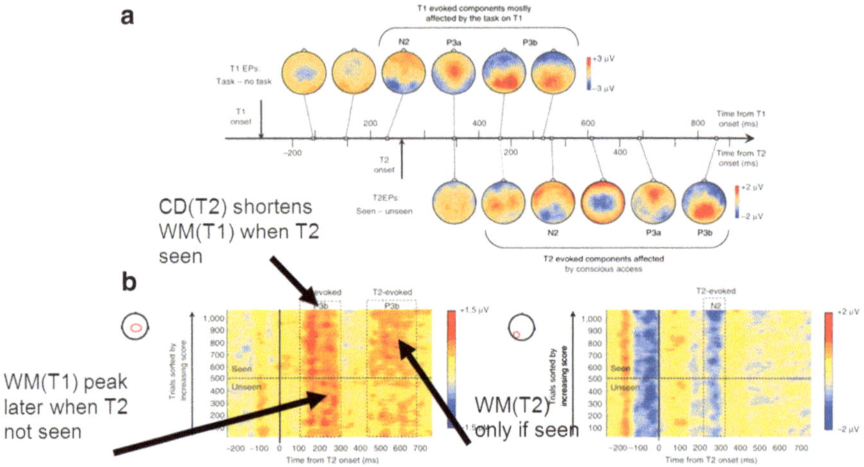

Fig. 9.3 Support data from the attentional blink

10 per second. One particular letter target, denoted T1 may be, for example, a white X. The second target, denoted T2, could then be a white F. The attentional blink occurs if the time gap between T1 and T2 is about 270 ms, so after about two intermediate targets. In (a) of Fig. 9.3 the time line is shown for the experiment, with EEG patterns as observed from the timing of the presentation and observation of T1 (with T2 undetected); aligned underneath is the corresponding set of brain waves in response to the observation of T2.

The attentional blink occurs when a series of putative targets, such as letters, is briefly shone on a screen in front of a subject, at a rate of about 10 per second. One particular letter target, denoted T1, may be, for example, a white X. The second target, denoted T2, could then be a white F. The attentional blink occurs if the time gap between T1 and T2 is about 270 ms, so after about two intermediate targets. In (a) of figure 6 is shown the time line for the experiment, with EEG patterns as observed from the timing of the presentation and observation of T1 (with T2 undetected); aligned underneath is the corresponding set of brain waves in response to the observation of T2.

There are three sets of arrowed annotations in Fig. 9.3 of the activity shown for many tests as lined up horizontally from the onset of T2 (specified by the vertical axes in the figures (b) denoting the degree of awareness of T2 reported by the subjects). These annotations denote:

1. The presence at the N2 time for T2 of a shortening in time of the length of the duration of activity of the P3 signal for T1 observed in the parieto-central area (as indicated by the circle in the small brain at the right side of the figure);
2. The slightly later peaking of the T1 working memory P3 peak when T2 is not observed;

3. The presence of a P3 for T2 only when it is consciously seen (with greater than 50 % confidence), in the later peak to the left of the earlier T1 P3 peak in the first figure in (b) of Fig. 9.3.

There are numerous other features to be noted in the data reported in (Sergent et al. 2005), but the above points indicate the presence of a signal inhibiting and/or speeding up the processing of T1 into awareness as evidenced by the existence of the distortion of its P3 signal under the condition of T2 being observed, but not when T2 is unobserved. This is exactly as to be expected for the corollary discharge mechanism of attention movement, if it is used to inhibit distracters and/or amplify the target activity, as suggested in the original CODAM model (Taylor 2000, 2002a, b).

The resulting time line, with relevant brain sites where possible, now becomes (dropping the earlier feature analysis interval):

Final Time Line

$$0\ \text{(stimulus onset, early visual cortex)} \rightarrow 180-200\ \text{ms}\ \Big(\text{attention movement signal creation as N2pca in SPL, as reported in Hopf et al. 2000)} \rightarrow 220-330\ \text{ms (corollary discharge activity in parietal)} \rightarrow 350-1{,}000\ \text{ms or beyond until report (P3 and access of attended stimuli to VSTM, with the corollary discharge being involved in further manipulation, as associated with the SPCN}\Big) \tag{9.2}$$

9.2.7 *Support from RLIP*

The above experimental data is supported by more recently reported cue-based data in (Van der Stigchel et al. 2006; see also Munneke et al. 2008). The crucial component was observed by EEG during the response by subjects in a psychological paradigm involving covert detection of a target amongst distracters. The paradigm involved a prior arrow cue indicating not only the direction of a target but also that of a distracter. As the authors stated (p38) "...we found a reversed effect in the EDAN latency range" (in the post-stimulus time range of 220–330 ms) "in response to inhibitory cues: a positive deflection in particular over right parietal regions contralateral to the direction of the cue (which they denoted as RLIP)" The RLIP as observed by (Van der Stigchel et al. 2006) is shown below in Fig. 9.4.

The RLIP signature in the ERPs was interpreted by the authors as arising from inhibition of the distracter position. What is important here is that this RLIP activation must have partaken of the attention control movement generation system in SPL/IPS/FEF, as noted in the quote above taken from the authors. For the RLIP signal requires knowledge of which of the visual stimuli presented to a subject were distracters and which was the target to be attended and responded to (similar to the

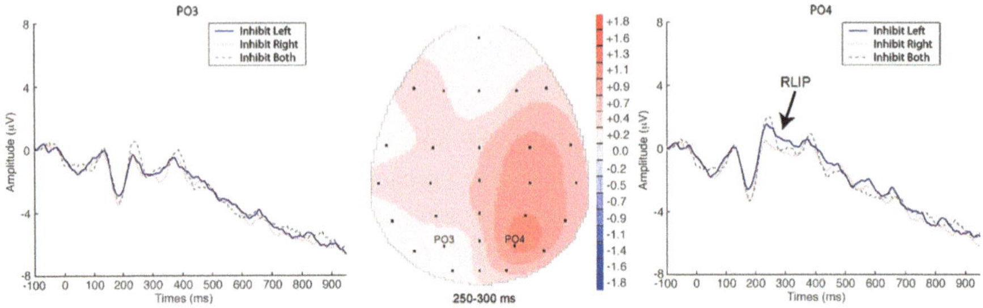

Fig. 9.4 The RLIP component

situation in the paradigm of R & J). The RLIP signal occurred at a still early stage in the information processing, just a little after we would expect the main attention feedback signal itself, but slightly later than the early N2pc observed in (Hopf et al. 2000) and reported above and in Fig. 9.1. This timing gives the RLIP signal time to use some components of the attention movement signal generated in SPL/IPS and observed in (Hopf et al. 2000) at 180–200 ms.

The left and right hand figures show the activations at the PO3 and PO4 electrodes, and in particular the RLIP signal they observed in the right hand figure. The central figure is of the scalp distribution of electrical activity associated with the set of EEG measurements in this paradigm, which involved a prior arrow cue indicating not only the direction of a target but also that of a distracter. RLIP = Right lateralized inhibition positivity (adapted from (Van der Stigchel et al. 2006) with permission).

We regard the data of Fig. 9.4 as further for proof of the existence of the attention movement corollary discharge signal. This is seen by the RLIP activity arising at an intermediate stage in the attention processing range, as just noted, and also occurring at a relatively high level in the processing (in parietal lobe). The RLIP is taken to correspond to the corollary discharge data from 220 to 330 ms in the target-based time line (1), with the cue now replacing the target as a stimulus used to control the movement of attention.

9.3 Further Data

As noted earlier, there are increasing numbers of components of ERP activity now being observed in subjects' brains during processing involved in attention-based paradigms. Some of this data involves cue-based ERP components (as in Van der Stigchel et al. 2006) as well as in target-based activity. The use of both of these two types of paradigms is argued by some to be unallowable (even termed 'neither permissible nor informative'). Yet if attention-movement-based signals arise from the subject moving attention to one hemisphere or to another by a cue observed before a target appearance, it would still seem legitimate to consider the manner of movement of such attention brought about by the cue. Such data is all grist to the attention mill, it can be argued. We will thus allow this data as part of our analysis, although with the proviso that it be noted as cue-based.

9.4 The Implications in Terms of CODAM

The dynamical mechanism whereby CODAM can be recognised as a neural brain-based model of consciousness has been considered in a number of the references on CODAM already cited above, and discussed more fully in Chaps. 6, 7, and 8. Thus we will only give a brief review of this possibility here. However we claim that such a model of consciousness is an important step forward in both the understanding of attention as the gateway to consciousness as well as for understanding the nature of consciousness itself.

The CODAM model assumes that there exists some module, denoted 'Owner' in Fig. 6.3 of Chap. 6, which allows for a brief holding in short term memory of the corollary discharge signal. As assumed for report by means of the well-supported visual short-term memory (VSTM), acting as a receptacle for report of the content of an incoming target stimulus, so it is assumed that the content of the corollary discharge short term memory would also be available for similar report, although possibly for a briefer time. However the presence of the owner signal of the corollary discharge of the attention movement signal gives this signal its content, which is that of 'ownership' of the about-to-arrive visual stimulus into report of that content. There can be no other nature of the experience generated in the corollary discharge short-term memory, since the activity there is not connected to lower level feature components enabling the stimulus activity to acquire content. Thus the owner activity is content free. But yet it possesses an experience of ownership due to the control it exerts over the access of the attended stimulus activity to its content report stage in the VSTM. Such control is assumed to consist of inhibition of possible distracters and amplification of the site for activation of the code for the attended stimulus.

We note here that amongst the various signals (N2pc, RLIP, SPCN etc) mentioned so far, the SPCN might be considered as the most crucial evidence to provide a neural basis for the ownership signature just mentioned, since it is based in a working memory system already. Thus the continued activity taking place in the SPCN as detected in Fig. 9.2, and certainly occurring over several hundred milliseconds pre-report, could provide a crucial neural infrastructure for the ownership component of the inner self. It may be conjectured that the length of time needed for disentangling target from distracter could be correlated with any increased sense of ownership of the resulting reported target identity in the R & J paradigm. However such analysis requires careful psychological experiments to be performed in which subjects are asked to rate their levels of ownership of the reported signal. If it were possible for subjects to sense this ownership, then in the R & J paradigm it is to be expected that the ownership experience would be expected to have a greatest strength for backward masking, be present to a lesser degree for forward masking and have least strength in the no mask case. This is a prediction which, however, may need great care to test in terms of subject response and its believability.

More generally, the implication of this content-free but owned experience is that it can tentatively be identified with the 'inner self' of Western phenomenology of Husserl, Merleau-Ponty, Sartre and many other philosophers. Although these all had different detailed takes on the inner or pre-reflective self they were all united as to its existence (Zahavi 2005). The inner self has since been teased out more fully in detailed studies by (Zahavi 2005) and by continued work on the process of loss of this content-free experience in schizophrenia (Sass and Parnas 2003; Taylor 2011).

As originally proposed by Husserl (Sokolowski 2000) there is a specific timing sequence for the emergence of a conscious experience of content. This was supposed to be in three stages:

$$\text{Pretention} \rightarrow \text{Primal Impression} \rightarrow \text{Protention}$$

Each of these three stages was distinct: pretention arose at the early stage of the consciousness creation, the primal impression was that of the content of the attended stimulus, and protention involved a buffered memory of the experience, as described already in the book.

We have earlier in Chaps. 6 and 7 (and originally in Taylor 2002a, b) described in some detail how CODAM can explain these three temporal segments of the emergence of consciousness. Pretention is to be considered as the stages and associated experience involved in the creation of the attention feedback signal, the related corollary discharge activity and attention amplification of visual cortical activity representing the attended stimulus. The primal impression is the emergence of the amplified attended stimulus activity onto its buffer working memory for general report round the brain. Finally the protention period involves the continued but decaying activity on the buffer working memory site, the VSTM. Such a division of the dynamic activities in CODAM is a natural one, and fits nicely with the results of the experiential explorations of Husserl and his colleagues.

We can view differently the temporal flow of experience from the above three components of the sequence by denoting the early processing, originally under the heading of 'pretention', now to be put under the different heading of 'ownership'. Such ownership involves the detailed control processes (inhibition and amplification) proposed for the corollary discharge signal and claimed above to have been observed in various paradigms (Hopf et al. 2000; Van der Stigchel et al. 2006; Sergent et al. 2005; Robitaille and Jolicoeur 2006).

In a manner similar to that in which the external world attains a constant form by means of the eye-movement corollary discharge (Berman et al. 2007), so we can expect that the ownership experience, that of the 'I', can be kept constant by means of the attention corollary discharge signal. This would thereby lead to what can be termed the 'Constant I', which is as directly experienced by each of us as we move through the world. The exact mechanism for this constancy is still unclear in the case of the external vision of the world (see Cavanagh et al. 2010 for a very recent discussion on this). In a similar manner we cannot conclude on a specific mechanism for the constant 'I'. However we can expect there to be a close analogy between these two mechanisms from the analogy of the existence of the two corollary discharge mechanisms, the first for retinal movement and the second for attention movement.

There is still a large gap in our analysis of experimental support for the existence of a corollary discharge as both existing and being at the basis of consciousness. There are many other situations in which the focus of attention is moved other than that analyzed by Robitaille and Jolicoeur (2006). We have noted that there are a few other paradigms, which we mentioned above in Sect. 9.3; these also give a hint of the presence of a corollary discharge of the attention movement control signal. However these are still a very small proportion of all events in which the focus of attention is moved. We have not shown that such a corollary discharge is also present in all of these cases. We still have to look at the case when attention is held steady. We could hypothesise that there is then a constant refreshing of the movement control system, although that would have also to be carefully investigated. So there is still much to be done to prove that the CODAM model is suitably secure as being at the basis of the creation of consciousness.

9.5 Conclusions

Initial experimental evidence for the existence of a corollary discharge signal for the movement of attention, basic to the CODAM model, is presented from EEG and MEG data. This model may then be used to help explain certain aspects of conscious experience as noted by the school of Western phenomenologists (Husserl, Sartre, Merleau-Ponty, etc), in particular the existence of the inner or pre-reflective self, or 'I', and the associated temporal flow of consciousness. The resulting CODAM corollary discharge model of attention control is shown to provide both a possible explanation of the ownership component of consciousness as identified with the pre-reflective self, as well as hinting at explaining the constant nature of this self through various conscious experiences – as the 'Constant I'. However this avenue requires considerable further research to become established.

Finally there are clearly a number of outstanding points in our attempt to identify any corollary discharge of an attention movement control signal. In particular how do we know that the signal discussed is not part of the feedback attention control signal itself, and not just a copy of it? We have taken the dividing line of the feedback signal to be the hierarchy of visual sites both by the dorsal and ventral routes. But how far up in the visual processing hierarchy do these signals go? Our answer is that they should not go into parietal lobe modules for vision. That may need revision after more careful analysis, as along the lines of (Mehta et al. 2000). That would allow both the initial visual input signal and the feedback attention signal to be observed, so identifying the modules under direct attention control. Until such more detailed analysis is achieved, any conclusions we reach with present data can only be considered as a preliminary view of the overall attention control system. We also need to check the many other cases of the movement of attention focus, as well as when attention is held fixed, to justify that such a corollary discharge is always at the basis of any conscious experience.

Acknowledgements The author would like to thank Dr N Fragopanagos for discussions and simulations across a range of attention paradigms, and Drs Hopf, Sergent, Robitaille & Jolicoeur and Stigchel for granting permission to use their data. The helpful comments from three reviewers are gratefully acknowledged, as is partial financial support from the EU under the DARWIN Project.

References

Berman RA, Heiser LM, Dunn CA, Saunders RC, Colby CL (2007) Dynamic circuitry for updating spatial representations. III. From neurons to behavior. J Neurophysiol 98(1):105–121

Bressler SL, Tang W, Sylvester CM, Shulman GL, Corbetta M (2008) Top-down control of human visual cortex by frontal and parietal cortex in anticipatory visual spatial attention. J Neurosci 28(40):10056–10061

Cavanagh P, Hunt AR, Afraz A, Rolfs M (2010) Visual stability based on remapping of attention pointers. Trends Cogn Sci 14(4):147–153

Corbetta M, Patel G, Shulman G (2008) The reorienting system of the human brain: from environment to theory of mind. Neuron 58:306–324

Desmurget M, Grafton S (2000) Forward modelling allows feedback control for fast reaching movements. Trends Cogn Sci 4:423–431

Fragopanagos N, Taylor JG (2007) Resolving some confusions over attention and consciousness. Neural Netw 20(9):993–1003, Special Issue on Brain and Attention (to appear)

Fragopanagos N, Kockelkoren S, Taylor JG (2005) A neurodynamic model of the attentional blink. Cogn Brain Res 24:568–586

Gregoriou GG, Gotts SJ, Zhou H, Desimone R (2009) High-frequency, long-range coupling between prefrontal and visual cortex during attention. Science 324:1207–1210

Hickey C, DiLollo V, McDonald JJ (2009) Electrophysiological indices of target and distractor processing in visual search. J Cogn Neurosci 21(4):760–775

Hillmire MR, Mounts JRW, Parks NA, Corballis PM (2010) Event related potentials differentiate effects of salience and space in biased competition for visual representation. PLoS One 5(9):e12677. doi:10.1371/journal.pone.0012677

Hopf J-M, Luck SJ, Girelloi M, Tillman H, Mangun GR, Scheich H, Heinze H-J (2000) Neural sources of focussed attention in visual search. Cereb Cortex 10:1233–1241

Hopf J-M, Boelmans K, Schoenfeld AM, Heinze H-J, Luck SJ (2002a) How does attention attenuate target-distractor interference in vision? Evidence from magnetoencephalography recordings. Cogn Brain Res 15:17–29

Hopf J-M, Vogel EK, Woodman G, Heinze HJ, Luck SJ (2002b) Localizing discrimination processes in time and space. J Neurophysiol 88:2088–2095

Hopf J-M, Boelmans K, Schoenfeld MA, Luck SJ, Heinze H-J (2004) Attention to features precedes attention to locations in visual search: evidence from electromagnetic responses in humans. J Neurosci 24(8):1822–1832

Kiss M, Van Velzen J, Eimer M (2008) The N2pc component and its links to attention shifts and spatially selective visual processing. Psychophysiology 45:240–249

Korsten N, Fragopanagos N, Hartley M, Taylor N, Taylor JG (2006) Attention as a controller. Neural Netw 19(9):1408–1421
Luck SJ, Hillyard SA (1994a) Spatial filtering during visual search. Evidence from human electrophysiology. J Exp Psychol Hum Percept Perform 20:1000–1014
Luck SJ, Hillyard SA (1994b) Electrophysiological correlates of feature analysis during visual search. Psychophysiology 31:291–308
Mehta AD, Ulbert I, Schroeder CE (2000) Intermodal attentional selection in monkeys I: distribution and timing of effects across visual areas. Cereb Cortex 10(343):358
Munneke J, Van der Stigchel S, Theeuwes J (2008) Cueing the location of a distractor: an inhibitory mechanism of spatial attention? Acta Psychol (Amst) 129:101–107
Rizolatti G, Craighero L (2010) The premotor theory of attention. Scholarpedia 5(1):6311
Robitaille N, Jolicoeur P (2006) Fundamental properties of the N2pc as an index of spatial attention: effects of masking. Can J Exp Psychol 60(2):101–111
Robitaille N, Grimault S, Jolicoeur P (2009) Bilateral parietal and contralateral responses during maintenance of unilaterally-encoded objects in visual short-term memory: evidence from magnetoencephalography. Psychophysiology 46:1090–1099
Sass LA, Parnas J (2003) Schizophrenia, consciousness and the self. Schizophr Bull 29(3):427–444
Schenkluhn B, Ruff CC, Reinen K, Chambers CD (2008) Parietal stimulation decouples spatial and feature-based attention. J Neurosci 28(44):11106–11110
Sergent C, Baillet S, Dehaene S (2005) Timing of the brain events underlying access to consciousness during the attentional blink. Nat Neurosci 8:1391–1400
Sokolowski R (2000) Introduction to phenomenology. Cambridge University Press, Cambridge
Sridharan D, Levitin DJ, Chafe CH, Berger J, Menon V (2007) Neural dynamics of event segmentation in music: converging evidence for dissociable ventral and dorsal networks. Neuron 55:521–532
Taylor JG (1999) Race for consciousness. Bradford Book, Cambridge, MA
Taylor JG (2000) Attentional movement: the control basis for consciousness. Soc Neurosci Abstr 26, 2231#839.3
Taylor JG (2002a) Paying attention to consciousness. Trends Cogn Sci 6(5):206–210
Taylor JG (2002b) From matter to mind. J Consciousness Stud 6:3–22
Taylor JG (2007) CODAM: a model of attention leading to the creation of consciousness. Scholarpedia 2(11):1598
Taylor JG (2011) A neural model of the loss of self in schizophrenia. Schizophr Bull 37(6): 1229–1247
Taylor JG, Fragopanagos N (2003) Simulations of attention control models in sensory and motor paradigms. In: Wunsch II DC, Hasselmo M, Venayagamoorthy K, Wang D (ed) Proceedings of the international conference on artificial neural networks (ICANN 2003). IEEE Press, Los Angeles, pp 298–303
Taylor JG, Rogers M (2002) A control model of the movement of attention. Neural Netw 15:309–326
Van der Stigchel S, Heslenfeld DJ, Theeuwes J (2006) An ERP study of preparatory and inhibitory mechanisms in a cued saccade task. Brain Res 1105:32–45
Vogel EK, Machizawa MG (2004) Neural activity predicts individual differences in visual working memory capacity. Nature 458:748–751
Wolpert DM, Ghahramani Z (2000) Computational principles of movement neuroscience. Nat Neurosci 3:1212–1217
Wurtz RH, Sommer MA (2004) Identifying corollary discharges for movement in the primate brain. Prog Brain Res 144:47–60
Zahavi D (2005) Subjectivity & selfhood. MIT Press, Cambridge, MA

Chapter 10
Understanding the Mental Disease of Schizophrenia

10.1 Introduction

There are numerous mental diseases that yet require deep investigation in terms of the alteration of the mental state of the patient due to the disease onset. Of course there are other aspects of mental diseases which obviously urgently need exploration, such as cause, cure, etc. However it is only appropriate to consider the alteration of the patient's mental state in this book. And of all the mental diseases it is schizophrenia which is most relevant in looking for the mental disease with greatest effect on the subject's mental state. Numerous researchers have pointed out over the last decades that there is a loss of the sense of the inner self in schizophrenia. In particular the illuminating paper of (Sass and Parnas 2003) gives an underpinning explanation of the disease along these lines in each of its three manifestations, with positive, negative or disordered symptoms (see also Cermolacce et al. 2007; Parnas et al. 2003, 2005b; Sass and Parnas 2007; Sass 1992). The crucial component of the analysis of these researchers is that of various disturbances in ipseity (the ongoing sense of 'being there' accompanying all conscious experience) that can occur for a sufferer, giving a framework with which to understand the disease. Such analyses of schizophrenia in terms of distortions of the self go back much earlier (Berze 1914; MinkowskiLa 1927; Blankenburg 2001; Kimura 1992; Sass 2001). However the more recent work has become more precise and embracing in terms of seeing most forms of schizophrenia as arising from such distortions. It also provides new ways of looking at and diagnosing the disease.

In this tenth chapter we propose to move the whole analysis closer into the brain itself by means of the CODAM neural network model of consciousness (Taylor 2000, 2007) (where we have already noted in several places that CODAM is an acronym for the 'COrollary Discharge of Attention Movement' model of consciousness creation, as described in Chap. 6). This allows both a mechanism to be formulated as to the basic brain-based cause of schizophrenia (with varieties

J.G. Taylor, *Solving the Mind-Body Problem by the CODAM Neural Model of Consciousness?*, Springer Series in Cognitive and Neural Systems 9,
DOI 10.1007/978-94-007-7645-6_10,

of this cause correlated with the three main forms of it) as well as open up possible lines of research to be followed to help ameliorate the attention defects exposed by this approach.

The increased understanding of schizophrenia as arising from distortions of the self have provided important insights into schizophrenia, as noted above. Thus as stated in (Sass and Parnas 2003) "..., schizophrenia is a disorder involving subtle but pervasive and persistent aspects of subjective experience". Furthermore, these authors state that (Sass and Parnas 2003, p. 428) "Schizophrenia we propose, is a disorder or more specifically an ipseity disturbance, in which one finds certain characteristic distortions of the act of awareness". Ipseity denotes the inner self, to be regarded as the pre-reflective self of Western phenomenology (Sass and Parnas 2003)· . In this book we identify ipseity with the inner self, and that with what has been termed the pre-reflective self (Zahavi 2005), although there may be subtle differences between them according to different definitions in Western phenomenology (Zahavi 2005) ; these will be ignored here (as earlier in the book). We use the definition of ipseity of (Sass and Parnas 2003, p. 428) "... the experiential sense of being a vital and self-coinciding subject of experience or first person perspective on the world". It arises from a part of the self which is devoid of the components of the reflective self, which itself is composed of those characteristics of the self that can be obtained by reflecting, such as whether or not one has a beard or is impatient, and so forth. The pre-reflective self appears instead as content-free, and its existence provides what appears to be a centre of gravity in which the ownership of one's experiences is gathered.

The more detailed analysis in (Sass and Parnas 2003), together with further studies of these authors and their colleagues on various aspects of possible breakdown of ipseity (Parnas et al. 2003, 2005b; Sass and Parnas 2007; Sass 1992), give an impressive attack on the gross functionality of schizophrenia and allow it to be considered in a revealing light with respect to the nature of the sufferer's experience.

In spite of the progress thereby made there is still the difficulty of a lack of fundamental understanding of the crucial component at issue: ipseity, or the inner self. Thus the insights of the Western philosophers of the past and of the more recent present (Zahavi 2005) have not led to any understanding of the manner in which the brain can help create the inner self [but see Sass 1992, chap 7, appendix; (Hemsley 1998) ; both being about the hippocampus-based "comparator" function approach, which cannot be correct, as the example of the patient HM without a hippocampus but seemingly very conscious demonstrates] nor of the brain's role in mediating how that inner self interacts with the stimulus representations of the outside world.

This interaction of the inner self with the outside world, however, plays a crucial role in the new understanding gained for schizophrenia. The distortions observed in schizophrenic subjects are concerned in many cases with the manner in which the outside world becomes more difficult to be dealt with by the sufferers. Thus in the case of Maria (Cermolacce et al. 2007), she is reported as saying that "She was never able to [...] immerse herself in the world because an invisible barrier prevented her full presence" (Cermolacce et al. 2007, p. 706). Similar accounts of this difficulty of dealing with the outside world are reported by many other schizophrenics. But how does the interaction between the inner self and the representations of the external world in the brain become distorted in schizophrenia? Such distortion plays a very important role in the disease but it has not yet been possible to determine how they are caused. This is due to the absence of any neural model of how ipseity or the inner self is itself represented in the brain. In particular for schizophrenia, as pointed out in (Sass and Parnas 2003), the main problem in understanding schizophrenia is as to how the breakdown of inner consciousness leads, in a sufferer, to the two major components pinpointed in (Sass and Parnas 2003):

1. Hyperreflexivity, in which there is an exaggerated form of attention being paid to self features as if they were external objects. As specified in (Sass and Parnas 2003, p. 428) hyperreflexivity denotes: "... forms of exaggerated self-consciousness in which a subject or agent experiences itself, or what would be normally inhibited as an aspect or feature of itself, as a kind of external object". It seems likely that this may result from a breakdown of some form of inhibition to prevent the unexpected awareness of these self features from breaking through into consciousness;
2. Reduction of what is termed 'self-affection', in which there is a diminished implicit sense of 'being there', so of self-presence. The term 'self-affection' is not related to liking or other emotion, but refers to the manner in which the sense of subjectivity can affect itself and acknowledge its own presence. This, it was claimed in (Sass and Parnas 2003) was reduced as an important basic component of experience in schizophrenics.

The first of these components can be recognized as part of the larger spectrum of 'loss of common sense', and will be treated as such later in the chapter in terms of the neural model, CODAM, of consciousness. This allows building a bridge from the brain of the schizophrenic to their inner experience. In particular this bridge will help pinpoint how the damage to the schizophrenic's experience can be seen to arise from a concomitant damage to their attention control system in the brain, as a part of damage to the CODAM model itself.

The second component will also be considered in terms of damages to CODAM leading to the resultant altered experience of the schizophrenic. I propose therefore to use this model since it helps understand the nature of the inner self and thereby the two important components recognized in (Sass and Parnas 2003) as basic to the schizophrenic experience.

In (Sass and Parnas 2003) it was pointed out that these two distortions – of hyperreflexivity and of self-affection – are in fact complementary aspects of a single underlying component, that of the inner self or ipseity. This component (ipseity) and the possible damages to its functionality is also the centre of our approach to schizophrenia through the CODAM model. In our view CODAM implements a source of the experience of ipseity itself.

It should be added that there are two forms of hyperreflexivity noted in (Sass and Parnas 2007) , termed 'reflective' and 'operative'; the former is of a largely willed or voluntary kind, the latter is unwilled or involuntary. It is this latter more basic form of hyperreflexivity that is the primary focus in (Sass and Parnas 2003), and with which we are concerned in this chapter. We note, then, that the ipseity component recognised in the CODAM model to be discussed shortly is not a willful or volitional process, and that distortions of such a neural system will be of the operative hyperreflexive type.

One important problem presently faced about schizophrenia, as seen from this vantage point of a breakdown of common sense about the world and of one's presence in it, is that of comprehending how to move this understanding further into the brain and thereby how the disease could be ameliorated. Much success has been gained recently by new drugs for the disease as well as by new behavioral treatments (such as cognitive behaviour therapy (CBT) and attention/cognitive training schemes). However such drug-based or CBT-based advances do not seem to be much related to the deeper understanding arising from the appreciation that ipseity distortion plays a crucial role across the whole range of schizophrenic symptoms.

One of the threads of this chapter is that we can begin to bridge this gap between these two approaches – that by drug and behavioral treatments and that of ipseity distortion – by creating a specific brain-based model. This model would take account of the former treatment and activity-based approaches by including a neural network model of attention and its related cognitive processing powers (with an associated biochemical basis). At the same time its neural activity would also be interpretable in terms of various stages that arise in the creation of conscious experience. Such a model of consciousness is obviously controversial, but needs to be attempted to make progress towards a more unified view of schizophrenia.

Over the last decade a model of consciousness as based on the brain has been developed (Taylor 2000, 2007), and was described in Chap. 6. It was felt appropriate to give a brief reprise on CODAM here in order to have CODAM fresh in the reader's mind. The model uses attention as its basis; importantly it extends the standard ballistic control model of attention to include important use of a corollary discharge of the attention movement control signal. This is proposed to lead to a signal of ownership of the about-to-arrive conscious experience. The CODAM model thereby leads to a description of consciousness fully consistent with that arising from the basic analyses of Western phenomenology (Sokolowski 2000). In particular we have already discussed the 'protention –primal impression – retention' main sequence of events for consciousness discovered through phenomenology, as explained by the attention-based CODAM model in terms of the necessary dynamics of the processing of attention to support object-based neural activity accessing buffer working memories for report (Taylor 2010). By way of further explanation, we add that these three fundamental components of experience are defined as follows: protention is equal to the preparatory activity leading to consciousness, the primal impression is defined as the actual first moment of conscious experience of the external stimulus, retention is the subsequent decaying memory of this conscious experience, that may last some seconds after the primal impression occurred.

The purpose of this chapter is firstly to explore possible distortions in the process of creation of the inner self through CODAM. Secondly it is then shown how such distortions as are provided by the descriptions of the inner experience of schizophrenics can be explained by certain distortions specifically in the process of inner self creation in the brain-based CODAM model, and hence in the brain.

Attention is well known to be disturbed in schizophrenia. The attentional blink is a paradigm in which a rapid sequence of stimuli, occurring at say 10 Hz, is presented to a subject who has to detect first one target and then a second. The subject has the greatest difficulty if the time lag between the two stimuli is about 270 ms, the so-called 'attentional blink'. The paradigm is well known to be sensitive to various components of attention and has been shown to be increased in schizophrenics as compared to controls (Wynn et al. 2006). This implies that schizophrenics have a lower level of attention control of stimuli in the rapid serial visual presentation task, to which they are exposed in the attentional blink, than do controls (Wynn et al. 2006). Other paradigms also indicate the lower level of attention control (and related cognitive control) experienced by schizophrenics (Kopp 2007; Uhlhaas and Mishara 2007) as compared to normals.

Attention is a component of information processing for which there exists some evidence that its strengthening (by suitable exercises) can go towards ameliorating the symptoms of schizophrenia. This may help explain the successes gained by schizophrenics involved in the various cognitive training programs (Velligan et al. 2006; Kopp 2007).

The approach being taken in this chapter is that attention is a control system somewhat similar to that of motor control in the brain. This has been advocated by the premotor theory of attention (Rizolatti et al. 1994), and allows the application to attention of motor control ideas (Desmurget and Grafton 2000) (for which there is increasing experimental support). Disturbances in motor control and especially in agency (where agency is the knowledge of who is performing an action on one's body: oneself or another) have been put forward as basic to the difficulties of schizophrenics (Frith 1992). The extension of the analysis of similar defects to the attention control system thereby allows investigation of the conscious experience of the schizophrenic if one accepts that attention functions as the gateway to consciousness and the inner self (strongly supported by the neuroscience community), and as discussed in depth in Chap. 3. Such a shift also allows avoidance of the strictures in (Gallagher 2000, 2004; Cermolacce et al. 2007) against the motor control approach to schizophrenia (and especially that of distorted agency). For attention and motor control are seen to be somewhat divorced in the brain (Rushworth et al. 2001). More importantly the CODAM attention control model of (Taylor 2000, 2007) Chap. 6 avoids the difficulty of having to need the intention of a thought before experiencing it; this avoidance is obtained through the attention circuitry employed in the CODAM model. This will be investigated shortly in more detail.

Other approaches to a brain-based understanding of schizophrenia have been attempted. Thus a recent paper in this journal (Stephan et al. 2009) proposes that distorted synaptic learning processes are at the root of schizophrenic symptoms, leading to poor monitoring of actions. Our approach is more specific than this as to brain circuitry involved, but can accommodate the modifications of on-line learning suggested by these authors. However our proposal attempts to relate more closely to experiential modifications that changes in brain circuitry bring about than does the model in (Frith 1992). Moreover the alterations of the reported inner experience of the schizophrenic are regarded here as crucial clues as to what modifications of underlying brain circuitry need occur in order to bring about the various reported schizophrenic experiences.

A brain-based approach must also take proper account of modifications in brain connectivity and grey cell count as observed experimentally in schizophrenics. A paper from Thompson and colleagues (Thompson et al. 2001) has looked carefully at the developmental changes in the brains of early-onset schizophrenics as compared to normal adolescents. Their discovery of a wave of excessive grey-matter loss as compared to controls, starting in the parietal lobes, is an important guide to any relation to overall circuitry assumed to be damaged or otherwise modified in such schizophrenics. Since much of the underlying CODAM brain circuitry is now understood (Corbetta et al. 2008) it is possible to use such graded loss observed (Thompson et al. 2001) to relate and understand experiential changes endured by the young developing schizophrenics.

10.2 Schizophrenia Explained by CODAM?

We attack the problem of explaining some of the symptoms of schizophrenia through the attention copy CODAM model of Chap. 6. We consider each of the three kinds of symptoms sequentially, using the discussion of (Sass and Parnas 2003). We also include in our considerations the prodromal symptoms considered in (Sass and Parnas 2003), regarding this as indexing a precursor stage of the disease.

10.2.1 Prodromal Symptoms

Such symptoms have been carefully discussed in (Sass and Parnas 2003), where the paucity of data was noted but also that there were results from follow-back studies of the early experiences of schizophrenics. In particular the study of (Klosterkotter et al. 2001), confirming earlier studies, showed that in the prodromal stage there are a multitude of anomalous experiences such as varieties of depersonalization, disturbances of the stream of consciousness and distorted bodily experiences. As noted in (Sass and Parnas 2003) the patient complains of a profound change but cannot easily describe it; complaints can be from 'I don't feel myself' or 'I am not myself' to 'I am losing contact with myself' or 'I am becoming a monster'. A patient will even say that 'My I-feeling is diminished' or 'My I is disappearing for me' (Sass and Parnas 2003, p. 438).

Such symptoms could arise from a CODAM-type of attention control by a reduction of the activity on the corollary discharge (attention copy) signal buffer posited in CODAM to be the source of the experience of the 'inner self' or 'owner'. It is this component in CODAM that is supposed to provide the sense of ownership of the forthcoming phenomenological experience of content – the red of the rose, the bouquet of the wine, the flavor of the steak. If there is reduced activity on the corollary discharge buffer then, by CODAM, this ownership will itself be reduced and the experience of 'losing contact with myself' would have a physiological basis. Thus the hypothesis fitting many of the prodromal symptoms is that they are caused by incipient reduction in the activity of the corollary discharge buffer.

10.2.2 Positive Symptoms

These symptoms, especially the first rank symptoms, are (as noted in [1)] defined by a 'kind of diminished self-affection'. This is described by a reduction of the sense of inhabiting one's own thoughts, feelings, actions etc. All have become alien, to some degree, even to the point of regarding them as under the control of an alien force or other person. A sufferer may comment that they 'only feel half there'.

Such a sense of loss can be ascribed, by the CODAM model, to the reduction of the normal effect of the corollary discharge/ attention copy signal in aiding the brain's attended stimulus representation attain awareness. In the positive symptom schizophrenic case, then, the attended stimulus is thereby reduced in its activity when it finally achieves consciousness. There will therefore be expected to be a lowering in this process of the involvement of the inner self with the resulting sensory buffer activity. Extending this to all lower level brain activity to which attention can be paid leads to the diminished self-affection cited by various writers (Schnieder 1959; Mellor 1970).

We note that reduced activity on the corollary discharge buffer has already been introduced for the prodromal symptoms. Here we go beyond the lowered value of corollary discharge activity on its buffer by considering in more detail the lower value of the degree of inhibition produced on the sensory input buffer by the output of this corollary discharge. This reduction will successively let in more and more distracters as the level of inhibition is reduced. But then inputs of which the subject is not normally conscious will arise in the subject's experience. This process may not be trivial in the case of inner speech causing hallucinations, as we now consider.

A great deal of discussion has been given over the last two decades about the thesis that damage to the motor control corollary discharge or some related motor control comparator system using such a signal in the schizophrenic brain is at the root cause of their hallucinations associated with inner speech, regarded as one of the important positive symptoms (Frith 1992). The agent (the subject) misattributes the origin of their inner speech to that of an alien. However the thesis has numerous problems, some noted earlier (Cermolacce et al. 2007; Gallagher 2004). In the CODAM approach the inner speech is still in the subject's stream of consciousness. The content of the inner speech will be caused by subconscious thought processes emerging into the subject's consciousness due to emotional salience or some other value they possess (such as unpleasant remarks about the subject themselves). This inner speech will thereby attain the sensory buffer for report to other brain sites. However the subject does not own such inner speech (this not being a question of agency but ownership). This paradoxical situation ('it's in my consciousness as being reportable but I don't own it') is a situation explicable by the general feature of reduction of the power of the corollary discharge activity in controlling access to consciousness of input to the sensory buffer. If there was no strong corollary discharge signal at the emergence of the inner speech onto the sensory buffer then there would be no ownership tag associated with it. Hallucinatory phenomena as part of the positive symptoms, would thus be interpreted by the subject as involving loss of ownership. There is still the question as to how the hallucinations or the inner speech can enter into consciousness if not boosted by the corollary discharge mechanism of CODAM described earlier. This must involve a complex dynamical situation with, say, emotionally valued inner speech signals bringing attention to focus on itself as signals at both unconscious and working memory level. In other words attention would be drawn not only to lower level but even working memory level brain activity. Thus fully-owned consciousness would then arise, with the associated ownership experience (from the corollary discharge), soon after the inner speech attains the auditory buffer Thus the hallucinatory signals, on the CODAM approach, would enter fully into the subject's consciousness by a two-stage process:

(a) initial access of the inner speech onto the auditory buffer (due to its valence and unaided and uninhibited by the corollary discharge system);
(b) the focus of attention then being drawn to this salient inner speech at various levels in the brain, thereby inserting it into the subject's 'owned' stream of consciousness.

Undoubtedly the experience of having a signal being reportable across one's brain (the component (a) above) followed shortly thereafter by its being acknowledged as in one's own consciousness would be disturbing, and could well be ascribed as being caused by an outside agent. We note finally that this is the example mentioned earlier as to an empirical breakdown of the 'immunity to error of misidentification of the first person pronoun'.

10.2.3 Negative Symptoms

These include poverty of speech, affective flattening, avolition, apathy and a general lack of notice of the outer world. However, as importantly noted in (Sass and Parnas 2003) some subjects report no deficit in affect or thinking, but did report being lost in their own thoughts and reducing their actions. Following (Blankenburg 2001) it would appear that for the central defect (more generally in schizophrenia but most specifically for the negative symptoms) there is a 'loss of natural self evidence – of common sense orientation to the world'.

We can begin to see in general how such a loss could occur in the case of damage to the attention copy model architecture of Fig. 6.3 or 6.4. If the attention copy signal were especially reduced in its distracter inhibition component, then otherwise prevented inputs would be allowed into consciousness. This is especially noted in schizophrenic cases of bodily processes becoming available to consciousness.

The poet and essayist Antonin Artaud is a famous case of this. He describes his consciousness as being invaded by bodily functions which would normally be outside consciousness. Thus he writes of "the limbo of a nightmare of bone and muscles, with the sensation of stomach functions snapping like a flag in the phosphorescence of the storm" and "images of bloody old cottons pulled out in the shape of arms and legs, images of distant and dislocated members" (Artaud and Hirschman 1965; Artaud 1976; Sass 2003). In particular in (Sass 2003) there is a lengthy and deep description of the writings of Artaud and how his experiences are explicable in the framework of loss of ipseity.

More generally the common sense schema (concerning use of a familiar external object often met with) which normally operates outside awareness, would now intrude on conscious processing. As a result the originally automatic schemata would be coded in such a manner they could not be carried out under conscious control. Thus the subject would have to 'start all over again' to learn how to handle these distracters and so over-learn them to regain their being automatic. So the subject would feel uncertain as to how to proceed, and certainly could not do so automatically and smoothly. This would therefore explain the 'loss of natural self-evidence' in many cases of schizophrenia. Attention would have to be brought to bear on the schemata involved (if they involve responses) or the sensations (if they are only at the level of perceptions). Altogether the patient would be expected to be considerably slowed down, if not completely flummoxed, by such apparitions in their consciousness.

It is necessary to take careful note of how what would otherwise be automatic schemata can enter consciousness. These schemata would normally have no connection to the sites of consciousness creation, here being taken as the buffer working memory sites dotted about various higher-order sites in the brain. However to become an automatic schema a sequence of actions and visual states, for example, must be over-learnt. Initially there will have been effortful, attended processing of this sequence, so of connections to the relevant buffer sites (hence the schema is in consciousness). As learning proceeds to an over-learnt situation, the possible internal models trained to hold any such schema become outside attention (which is not now needed to be applied to run off the schemata in an error-free manner). In other words the internal models can function outside consciousness. The manner this occurs may be, for example, by inhibition of connections to the relevant buffers otherwise granting consciousness to the states of the internal model. But the problem in schizophrenia is that such inhibition may fail, as we are positing it does for the inhibition (of distracters) arising from the attention copy signal. There will then no longer be efficient prevention of such distracters from impinging on the visual buffer. This leads to a loss of 'common sense' about the world.

10.2.4 Disorganization Symptoms

These occur along the lines expressed by their title: incoherence and pressure of speech, poverty of speech content, distractibility, tangentiality and derailment. It is clear that these would occur if the overall attention control system was itself breaking down, although this may occur in either a patchy or a sequential manner. If the latter were so then there will be a whole spectrum of increasingly disorganized symptoms. The ones mentioned above finally concluding in an almost total breakdown in the overall control guidance by the attention control signal generator. On the other hand a patchy process of breakdown of the attention control circuitry will not lead to such an ordered process of breakdown of behavior but be closer to the variation of symptoms in many reported cases. In either situation the nature of the temporality of the overall loss of attention control is of importance in understanding the mechanisms involved in causing the overall control breakdown.

The writer Artaud described his experiences of this state graphically (Fragopanagos et al. 2005, p. 294):

> This slackening, this confusion, this fragility ... correspond to an infinite number of new impressions and sensations, the most characteristic of which is a kind of disappearance or disintegration or collapse of first assumptions which even causes me to wonder why, for example, red (the color) is considered red and affects me as red, why a judgment affects me as a judgment and not as a pain, why I feel pain, and why this particular pain, which I feel without understanding it.

Yet again the loss of common sense about the world is evident in Artaud's experience, but also the plethora of impressions rushing in on him, so that he finds it difficult to concentrate on a single one. However he is driven to do so and thence to lose a common sense approach to the world, with reduced ability to inhibit the components normally processed automatically in association with the various stimuli in his sensory field, such as the affordances they provide and the processing schemata they automatically generate in relevant internal models.

That there is a loss of distracter inhibition in schizophrenia is supported by the observed increased attentional blink in schizophrenics, as we discussed earlier in the paper.

10.3 Relating CODAM to Brain Sites Involved in Schizophrenia

The simulation of the behavioral results described above in terms of the effects on the AB of schizophrenia give support to the CODAM model of consciousness in its application to schizophrenia. There is further support arising from a more detail approach using results on changes in the structure of the brain as schizophrenia progresses. We take here the sites observed by Thompson and colleagues (Thompson et al. 2001), which were mentioned earlier as revealing most dramatically an excessive loss of gray matter compared to controls in the brains of early-onset schizophrenics (followed by yearly MRI brain scans from age 13 to 18 years).

In particular the earliest loss was of parietal brain tissue that preceded puberty. It was found that "Second, parietal and motor cortices showed a severe early deficit with diffuse loss in other (but not temporal) cortical regions." (Thompson et al. 2001, p. 11653). Subsequent loss of gray matter was then observed to occur in a 'dynamic wave of progression' from parietal cortices into superior frontal, dorso-lateral prefrontal and temporal cortices. Moreover this dynamic wave was conjectured as being triggered partly by genetic and partly by environmental influences, from the known deficit in parietal lobes of adult patients relative to their genetically identical controls (for monozygotic discordant twins).

Important further features were also reported in the study in (Thompson et al. 2001): Faster loss in temporal cortices was correlated across patients with a more severe level of positive schizophrenic symptoms, whereas faster loss rates in frontal cortex was correlated with patients with more severe negative symptoms.

In the CODAM model we can attempt to relate these loss rates and symptoms with defects brought about by the disease. This had already been discussed in the previous section, where we considered the four classes of symptoms, being: prodromal, positive, negative and disorganization symptoms. These could be considered as arising from successively enlarged defects in the processing by the various modules in the CODAM model of Figs. 6.3 and 6.4. Such enlargement can be seen to correspond to the dynamic wave of degradation of grey matter in various parts of the brain noted above and described more fully in (Thompson et al. 2001).

In more detail, we see that the prodromal symptoms can be ascribed to the parietal grey matter degradation in the earliest pre-pubertal stage in the adolescents in that report, where we site the attention copy buffers as part of the parietally placed working memory buffer sites. These are especially important in the CODAM model as supporting both the attention control signal generator and the sensory input and corollary discharge buffers. Critically degradation in the last of these modules will be expected to cause reduced output from that site, so reduce the level of inhibition of distracters; this would lead to the loss of common sense as well as reduction in the sense of 'I' as noted earlier as occurring in prodromal symptoms. Associated motor problems also fit into these results.

The later degradations in the temporal lobes are correlated with positive schizophrenic symptoms (Thompson et al. 2001). In CODAM such degradation are expected to produce degrading effects on object representations (in the temporal lobe) as well as those on the TPJ, a crucial part of an extended attention controller (on inclusion of the SPL) in providing exogenous attention control (Corbetta et al. 2008). The first of these degradations (those in the temporal lobe) would be expected to cause unexpected and noisy object activations, and lead to unexpected hallucinations not ascribed to oneself, both in audition as well as in vision, as already described in more detail above. There may also be loss of affordance codes (in STS, for example), so explaining some of the motor control deficits in schizophrenia. The second (associated with loss of TPJ) would lead to loss of attention to the unexpected appearance of stimuli, so the slowing of responses. This may contribute to the sense of loss of the self, as noted earlier under positive symptoms, as well as to a slight delay in the experiencing of consciousness reported by several schizophrenics.

Later loss of grey matter occurring in the prefrontal cortex will be correlated with the development of negative symptoms (Thompson et al. 2001). This would correspond to loss of goals and the resulting attention guidance, and fits well with the related negative symptoms of avolition and apathy. For the reduction of goal activation would lead to less overall involvement of attention to the outside world.

Not only are goals held in prefrontal cortex but also decision making occurs as guided by activities there, especially in the anterior cingulate cortex, but also supported by activity in dorsolateral prefrontal cortex. Thus if these areas are damaged (due to excessive loss of grey matter) then not only are goals more difficult to store but also decisions are difficult to make, even if goals are available. Prefrontal symptoms are well known to involve loss of decision-making ability, so this is also expected to occur during disorganization symptoms.

Finally the consolidation of degradation across these areas of the cortex would be expected to lead to increased breakdown of increasing many of the control circuits of CODAM in the brain and so to less overall control of attention. This was noted earlier as basic to the disorganization symptoms. Thus these symptoms are to be regarded as arising from the general breakdown of the CODAM attention control model.

Thus there would be expected a gradual progression of damage in some schizophrenics of their attention control system and related experiences of consciousness. It would be important to explore this aspect further in order to determine if such progression is observable across patients or if, instead, schizophrenic breakdown of the attention control system is patchy. Both possibilities should be looked out for: some patients having patchy attention control system degeneration, some having a continuous and ever more-embracing loss of attention control. These differences between gradual degeneration and patchy loss of control structures in the brain might be due to differences of damage to the neuromodulatory systems underpinning the overall attention control system.

10.4 Biochemical Underpinning

An excess or deficit of various neuro-modulators has been suggested as a causative factor in schizophrenia: too low a level of dopamine (as part of the hypo-frontality observed by numbers of researchers in schizophrenia) or excessive dopamine sensitivity (as in the striatum (Howes et al. 2009)), serotonin, noradrenaline (conjectured as crucial in the exogenous attention circuitry in the brain (Corbetta et al. 2008)), excessive or reduced acetylcholine (the basis of normal attention control). Any or all of these breakdowns in correct levels of normal function may occur. Here we consider the proposition that acetylcholine may be of crucial importance in schizophrenia, because of its known essential character in attention control (as discussed more fully shortly). Other neuro-modulators may ultimately need to be considered (as in the case of noradrenaline mentioned above) but there is considerable controversy about how this neuro-modulator acts in a global manner, whilst the understanding of acetylcholine is understood somewhat better. We will also seriously consider dopamine, following (Fragopanagos and Taylor 2006, Private communication).

Attention is often regarded as a mechanism by which attended objects become perceptually more salient, akin to increasing their contrast. It was demonstrated by means of human psychophysics that attention can also be described as a mechanism that reduces contextual integration, thereby ensuring that task irrelevant information is prevented from influencing the processing of task relevant information (Roberts and Thiele 2008). It has been suggested above that in the CODAM model this reduction can occur by inhibition. This was partially verified by the results of (Shoemaker 1968) on the attentional blink, a result to which we have already referred.

To investigate possible neuronal bases of this phenomenon there have been various studies of the effects of attention on spatial integration by cells in V1 of the macaque monkey. In line with their psychophysical results, attention directed to parafoveal locations reduced spatial integration by reducing the summation area of V1 neurons. Effects of acetylcholine application and attention were largely similar, with acetylcholine reducing spatial integration by reducing the neuron's summation area. These data demonstrated that attention can alter perceptual and neuronal spatial integration, and that acetylcholine contributes to task dependent receptive field dynamics. It can also provide a basis for the possible inhibitory effects observed previously. These effects were supported further by results of (Herrero et al. 2008), in which single cell recordings were taken from the monkey's visual area V1 while the monkey was performing an endogenously controlled attention task. In such situations, attended visual cells (one for which its receptive field is being attended to) increased its firing rate. Importantly a similar increase of the cell's firing rate was observed by application of a low dose of acetylcholine to the cell.

However these results have only explored how attention affects low-level cells by feedback amplification that could be supported by further acetylcholine input. From many experiments it is known that cell amplification under attention occurs in higher levels of cortex. (see, for example, (Sarter et al. 2005b)). Moreover neurophysiological studies have demonstrated that increases in cholinergic transmission in sensory areas enhance the cortical processing of thalamic inputs. Cholinergic activity also suppresses the retrieval of internal associations, thereby further promoting sensory input processing. Behavioral studies document the role of cortical cholinergic inputs in attentional functions and capacities by demonstrating, for example, that the integrity of the cortical cholinergic input system is necessary for attentional performance, and that the activity of cortical cholinergic inputs is selectively enhanced during attentional performance (as already discussed for V1 neurons above). Other neuromodulators (specifically Noradrenaline, Serotonin and Dopamine) do not possess this property.

It has been hypothesized (Sarter et al. 2005a) that the cortical cholinergic input system generally acts to optimize the processing of signals in attention-demanding contexts. Such signals 'recruit', via activation of basal forebrain corticopetal cholinergic projections, the cortical attention systems and thereby amplify the processing of attention-demanding signals (termed 'signal-driven cholinergic modulation of detection'). The activity of corticopetal cholinergic projections is also modulated by direct prefrontal projections to the basal forebrain and, indirectly, to cholinergic terminals elsewhere in the cortex; thus, cortical cholinergic inputs are also involved in the mediation of top-down effects, such as the knowledge-based augmentation of detection of signals and the filtering of irrelevant information. Thus, depending on the quality of signals and task characteristics, cortical cholinergic activity reflects the combined effects of signal-driven and cognitive modulation of detection. This hypothesis begins to explain signal intensity or duration-dependent performance in attention tasks, the distinct effects of cortex-wide versus prefrontal cholinergic deafferentation on attention performance, and it generates specific predictions concerning cortical acetylcholine release in attention task-performing animals.

This general approach (Sarter et al. 2005a) has also been applied to obtain a better understanding of schizophrenia. In particular it has been proposed in (Sarter et al. 2005b) that there is a correlation between the increased sensitization in schizophrenic patients to dopamine, especially in the striatum (see the latest experimental support for this in (Howes et al. 2009)) and that for acetylcholine in the cortex. The reasoning behind this causal chain from DA sensitivity to acetylcholine dysregulation (and excessive input to the cortex) has been suggested in (Sarter et al. 2005b) as being due to the following causal chain: increased dopamine sensitivity in the nucleus accumbens, the resulting increased GABergic signal from there to the basal forebrain, resulting in an excessive cholinergic output from there to the cortex. This excess of acetylcholine in cortex causes a severe disruption to the attention control system as well as lower level sites where attention is targeted.

We thus arrive at the Sarter-Bruno model of cholinergic disregulation of cortex as the main form of neuro-modulatory disturbance leading to schizophrenia, misplacing that of the dopamine-based hypoactivity of prefrontal cortex. As the authors state "In general this scenario extends traditional models of schizophrenia that have focused on the DA system and represents the hypothesis that dysregulation of the cortical cholinergic input system is the primary mediator in the information processing impairments in schizophrenia, specifically of the attentional abnormalities that contribute to the manifestation of psychotic symptoms" (Sarter et al. 2005b, p. 119).

More specifically the features of attention deficit in schizophrenia, such as observed quantitatively in the attentional blink (Wynn et al. 2006) would appear to arise from reduced inhibition; earlier we argued that the important schizophrenic symptom of the loss of common sense arose directly from the reduced ability to prevent distracters. The authors state that 'the performance of animals with an abnormally reactive cortical cholinergic input system is characterized by increases in the number of false alarms, given by the number of claims for targets present in non-signal trials' (Sarter et al. 2005a). They conclude that in such animals, 'attentional significance would be attributed to irrelevant or normally unattended stimuli' (Sarter et al. 2005a, p. 106). Furthermore they suggest that antipsychotic drugs work so as to reduce the dopamine load and hence, by a suitable biochemical causal chain, that of oversupply of acetylcholine. It is in that manner that an initially counterintuitive modulation (excess acetylcholine leads to attention dysfunction, whereas previously we noted how acetylcholine added to neurons leads to increased neural response) can be understood: in such a situation: thalamic (externally-based) inputs are emphasized, with correlated reduction of associative or top-down control.

An important aspect of the above explanation of the action of excessive acetylcholine in more detail is that of the separation between the action of acetylcholine on thalamic and on associative inputs. The former, it has been found experimentally, are differentially amplified, the latter suppressed by acetylcholine inputs. Moreover the former of these occurs through nicotinic receptors, the latter through muscarinic ones. This helps explain the effects observed in V1 mentioned above through a reduction of the lateral (associative) inputs to a cell as compared to the direct thalamic inputs. But then these latter would be increased in number, and hence there would be reduced prevention of them entering cortex.

It is in this way that it is possible to reconcile the effects of increased cortical acetylcholine with the 'loss of common sense' in schizophrenia. Thus lateral connections (regarded as inhibitory) in cortex would in general be decreased in effectiveness by the excess acetylcholine; the direct thalamic inputs would include distracters as well as targets. All of these latter inputs would be amplified by the excess acetylcholine, so leading to loss of common sense due to excess input information and increased difficulty of filtering out only the target stimulus input. That itself would be defined by some top-down signal (guided by a prefrontal goal state, for example). Such feedback would be regarded as associative, and hence reduced in effect on input signal filtering.

We conclude that the Sarter-Bruno model of acetylcholine excess in cortex in schizophrenics allows for an explanation of the important reduction of control by top-down feedback information in attention-based processing in cortex. In the CODAM model we see that an important component of such feedback involves the effect of signals from the corollary discharge buffer site onto the sensory buffer site. If that feedback has its associative effect reduced due to excessive acetylcholine then the sense of 'I am losing I' expressed by schizophrenics will be explained. The first person pronoun will not be misidentified but will find it increasingly difficult, as the acetylcholine level increases, to consider itself in control of the attended information processing. Ever more distracters will be attempting to attain and ultimately gain access to the visual working memory buffer. Hence the impression of loss of control over what is in consciousness will occur, and first rank symptoms will arise. As the acetylcholine level rises even further one can conjecture that the loss of effect of prefrontal goals increases to such an extent that firstly negative symptoms arise (due to inability to preserve a reasonable level of attention control) and finally disorganization symptoms will develop, with the attention control system almost completely out of control. However a linear progression from positive to negative to disorganization symptoms need not be present due to variations in levels of acetylcholine in different cortical sites as time develops, leading to the possibly patchy development of brain degradation mentioned earlier.

10.5 Implications for Diagnosis and Treatment of Schizophrenia

The main result of the above discussion is to give a joint brain architecture and neuromodulatory basis for the deficits of schizophrenia in its various symptoms across the range of prodromal, positive, negative and disorganization symptoms in terms of various breakdowns of the attention control system of the brain. In particular the CODAM model was employed to indicate how the several alterations in the experience of the inner self in schizophrenia could be understood in terms of specific degradations of control between modules in the CODAM model. Furthermore an explanation of how cortical acetylcholine levels can cause the known range of prodromal, positive, negative and disorganization symptoms was given on the basis of the Sarter-Bruno hypothesis, when employed in a general analysis of acetylcholine effects on experience as through the CODAM model.

A method being used in more detail to catch distortions of internal experience is, for example, the EASE questionnaire (Parnas et al. 2005a). This arose from experiential studies of several hundred schizophrenic patients, especially those in first admission. The EASE questionnaire was set up not only for the treating physicians but also to help the patient begin to describe what may be very strange and frightening experiences to them. In the process of making these experiences explicit it may be that patients begin to come to terms with them.

At the same time physicians can, aided by the responses to the questionnaire, begin to develop both drug and exercise treatments. Such exercise treatments are being used in terms especially of exercises for improving cognitive and attention-based processing (Velligan et al. 2006). This speaks directly to the CODAM model and the discussion given earlier as to how various breakdowns in the modules of CODAM help explain the various components of prodromal, positive, negative or disorganization symptoms. Further results coming from EASE responses to an ever increasing number of schizophrenics would be important to follow the inner experiences under the disease more closely.

10.6 Conclusions

Through the CODAM model we have presented an initial bridge between experience and the brain activities that produce that experience. CODAM is based on a model of attention control over the stimulus-driven activity in the brain. It is natural to point out that attention-controlled activity is not the only activity that the brain supports; automatic or subliminal activity is also of crucial importance in brain processing. Thus subliminal processing is recognized as the basis of creative processing. But also attention control is still important during the period of subliminal processing – such as in the period of creative thinking – due to the importance of such creative processing leading to an important subliminal thought and then attention directed back to the train of thought. Thus attention is still seen to play a crucial role throughout the various stages of brain processing. How schizophrenics can handle creativity tasks is unclear, although they would be expected to be worse than normals at such tasks due to their lower levels of control of attention. This could lead to a valuable research project to explore creative defects in schizophrenics as compared to normal subjects.

Attention is also brought into play when automatic motor control fails; we have not discussed here schizophrenics' difficulties with motor control at any length, considering it outside the remit of this paper. However we note that motor attention (intention) has been modeled along similar control lines to CODAM (which has been extended thereby to visuo-motor control systems (Hartley et al. 2008)). Thus motor control difficulties could also be pursued on a brain basis following that or similar models.

We then described in some detail how the inner self of consciousness could arise from a suitably precisely defined model of attention processing. In particular the corollary discharge of the attention movement control signal – the attention copy signal or CODAM model – was argued as providing a basis for the inner self. The CODAM model is crucially based on a component attention copy signal used to speed up access to consciousness. Moreover this copy signal can inhibit distracters which are important for each of us to keep out of consciousness, these distracters being normally processed at a subliminal level. Identification of the various components of the overall CODAM circuitry activity was also made with the protention – primal impression – retention sequence of Western phenomenology.

Turning to schizophrenia, this last aspect is basic to a possible understanding of the symptoms of loss of common sense in the world. We considered each of the four symptom classes of schizophrenia in turn: prodromal, positive, negative and finally disorganized symptoms. Each of these was explained in general terms through defects in the CODAM model brought about by damages to the variety of CODAM model modules. In general we found that initially defects were expected to arise in the attention copy buffer module and then spread successively across the CODAM model modules (although more patchy changes may also occur).

We then considered how dysfunction in the acetylcholine system could lead to the observed symptoms. The Sarter-Bruno hypothesis (that there is an increase in cortical acetylcholine above normal, driven originally by increased striatal dopamine) was shown to support the mechanisms considered briefly in the CODAM model. In particular the excess level of cortical acetylcholine was argued to be at the basis, through a CODAM model of top-down attention control, of modification of experiences as indicated by the prodromal, positive, negative and disorganization symptoms.

Finally we underlined the importance of attention-based exercise for possible amelioration of schizophrenic symptoms, in conjunction with medication. The CODAM model hypothesis implies that such amelioration would be mediated by reduction of cortical acetylcholine levels. Such reduction may also be attempted directly by suitably developed drugs.

We can thus summarize what has been presented as new in this chapter about schizophrenia as follows:

- A brain-based attention control architecture has been presented, through CODAM, to give an underlying framework by which the various components of consciousness can be decomposed into that of ipseity and the experience of content of attended stimuli;
- From the above model various of the symptoms of schizophrenia were explained as arising initially by damage to the corollary discharge component and ultimately by overall degradation of the functioning of the various modules of CODAM;
- The manner in which an excess of dopamine in the basal ganglia (Howes et al. 2009) can be seen as a causative feature in the breakdown of the functioning of the modules of CODAM due to excess of cortical acetylcholine;
- The crucial role played by the malfunction of the overall attention control system in schizophrenia (especially in disorganization symptoms), and the need for further exploration of how that might be reversed, say by drug-induced reduction of acetylcholine.
- A direct explanation of the positive symptoms of thought and speech insertion, by means of the breakdown of the ability of the attention corollary discharge signal to be involved in the accessing of inner speech and thought to the relevant sensory buffer (thereby avoiding the numerous difficulties of the explanation in (Frith 1992) in terms of their claimed damage to the motor control corollary discharge system).

At least four research avenues can be seen to arise from the brain-based model of schizophrenia presented in this chapter:

1. A more careful study of the manner in which attention is controlled in the normal brain, especially using brain imaging machines. Thus by fMRI and TMS (and also allied to EEG and MEG analyses) the flow of activation through the various modules of the control circuitry proposed under CODAM needs to be followed in more precision, especially pinpointing the attention copy signal components which are being used to provide the initial control aspects of buffer working memory access of attended stimuli.
2. An extension of this study to schizophrenics with a variety of symptoms, so as to clarify the degradation of the control circuitry as expected to occur from the results described in (Thompson et al. 2001): a particular approach to this would use the EASE questionnaire in order to clarify the inner experiences as completely as possible.

3. A more detailed study of the hypothesis introduced in (Sarter et al. 2005b), as described earlier in the paper, that an excess of acetylcholine is at the root of schizophrenia. This covers a broad range of components when applied to a model such as that of CODAM for consciousness, especially a detailed computational and experimental analysis of the amplification/inhibition brought about by a suitable level of acetylcholine into the attention control modules and the manner in which excess of acetylcholine causes attention control breakdown.
4. The need for a detailed study of exercises that could be developed to ameliorate the deficits arising in the corollary discharge component of the attention control system. This part of CODAM is particularly subtle, and its more complete understanding may be helped by specific attention exercises designed specifically to target it (say based on aspects of the attentional blink paradigm, or similar paradigms needing fast attention responses).

Acknowledgements The author would like to thank the EU for financial support associated with the EU Cognitive Systems MATHESIS project for partial support while this work was being completed, and Dr N Fragopanagos who performed the simulations of the damaged AB to model schizophrenic deficits in the AB.

References

Artaud A (1976) In: Sontag S (ed) Antonin Artaud: selected writings (trans: Weaver H). Farrar, Straus & Giroux, New York

Artaud A, Hirschman J (eds) (1965) Antonin Artaud anthology. City Lights Books, San Francisco

Berze J (1914) Die Primare Insuffizienz der Psychishcen Aktivitat: Ihr Wesen, ihre Erscheinungen and ihre Bedeutung als Grundstorungen der Dementia Praecox und des hypophrenen Uberhaupt. F Deutke, Leipzig

Blankenburg W (2001) First steps toward a psychopathology of "common sense" (trans: Mishara A) Philos Psychiatry Psychol 8:303–315

Cermolacce M, Naudin J, Parnas J (2007) The "minimal self" in psychopathology: re-examining the self-disorders in the schizophrenia spectrum. Conscious Cogn 16:703–714

Corbetta M, Patel G, Shulman GL (2008) The reorienting system of the human brain: from environment to theory of mind. Neuron 58:306–324

Desmurget M, Grafton S (2000) Forward modelling allows feedback control for fast reaching movements. Trends Cogn Sci 4(11):423–431

Fragopanagos N, Kockelkoren S, Taylor JG (2005) A neurodynamic model of the attentional blink. Cogn Brain Res 24:568–586

Frith CD (1992) The cognitive neuropsychology of schizophrenia. Erlbaum, Hove

Gallagher S (2000) Self-reference and schizophrenia: a cognitive model of immunity to error through misidentification. In: Zahavi D (ed) Exploring the self: philosophical and psychopathologica perspectives on self-experience. John Benjamin, Amsterdam/Philadelphia, pp 203–239

Gallagher S (2004) Neurocognitive models of schizophrenia: a cognitive model of immunity to error through misidentification. Psychopathology 37:8–19

Hartley M, Fagard J, Essaily R, Taylor JG (2008) Observational versus trial and error effects in a model of infant learning paradigms. In: Kurkova V, Kurkova-Pohlov V, Kotnik J (eds) Proceedings of the ICANN 2008, artificial neural networks. LNCS #5164. Springer, Berlin, Part II, pp 277–289

Hemsley DR (1998) Disruption of the "sense of self" in schizophrenia: potential links to disturbances of information processing. Br J Med Psychol 71:115–124
Herrero JL, Roberts MJ, Delicato LS, Gieselman MA, Dayan P, Their A (2008) Acetylcholine contributes through muscarinic receptors to attentional modulation in V1. Nature 454:1110–1114
Howes OD, Montgomery AJ, Asselin M-C et al (2009) Elevated striatal dopamine function linked to prodromal signs of schizophrenia. Arch Gen Psychiatry 66(1):13–20
Kimura B (1992) Ecrits de Psychopathologie Phenomenologique (trans: Boderlique). P.U.F, Paris
Klosterkotter J, Hellmich M, Steinmeyer EM, SchulzeLutter F (2001) Diagnosing schizophrenia in the initial prodromal phase. Arch Gen Psychiatry 58:158–164
Kopp B (2007) Mnemonic intrusions into working memory in psychometrically identified schizotypal individuals. J Behav Ther Exp Psychiatry 38(1):56–74
Mellor CS (1970) First rank symptoms of schizophrenia. Br J Psychiatry 117:15–23
Minkowski E (1927) La schizophrenia. Psychopathologie des shizoides et des schizophrenes. Payot, Paris
Parnas J, Handest P, Saebye D, Jansson L (2003) Anomalies of subjective experience in schizophrenia and psychotic bipolar illness. Acta Psychiatr Scand 108:126–133
Parnas J, Moller P, Thalbitzer J, Jansson L, Handest P, Zahavi D (2005a) EASE: Examination of Anomalous Self-Experience. Psychopathology 38:236–258
Parnas J, Handest P, Jannsson L, Saebye D (2005b) Anomalous subjective experiences among first admitted schizophrenia spectrum patients: empirical investigation. Psychopathology 38(5):259–267
Rizolatti G, Riggio L, Sheliga BM (1994) Space and selective attention. In: Umiltà C, Moscovitch M (eds) Attention and performance XV. Conscious and unconscious information processing. MIT Press, Cambridge, MA, pp 231–265
Roberts MJ, Thiele A (2008) Spatial integration and its moderation by attention and acetylcholine. Front Biosci 13:37–42
Rushworth MS, Johansen-Berg H, Gobel SMG, Devlan JT (2001) The left parietal lobe and premotor cortices: motor attention and selection. Neuroimage 20:S89–S100
Sarter M, Hasselmo ME, Bruno JP, Givens B (2005a) Unravelling the attentional function of cortical cholinergic inputs: interactions between signal driven and cognitive modulation of signal detection. Brain Res Rev 48:98–111
Sarter M, Nelson CL, Bruno JP (2005b) Cortical cholinergic transmission and cortical information processing in schizophrenia. Schizophr Bull 31(1):117–138
Sass L (1992) Madness and modernism. Harvard University Press, Cambridge, MA
Sass LA (2001) Self and world in schizophrenia: three classic approaches in phenomenological psychiatry. Philos Psychiatry Psychol 8:251–270
Sass L (2003) Negative symptoms, schizophrenia, and the self. Int J Psychol Psycholog Ther 3:153–180
Sass LA, Parnas J (2003) Schizophrenia, consciousness and the self. Schizophr Bull 29(3):427–444
Sass L, Parnas J (2007) Explaining schizophrenia: the relevance of phenomenology. In: Chung MC, Fulford KMW, Graham G (eds) Reconceiving schizophrenia. Oxford University Press, Oxford, pp 63–95
Schnieder K (1959) Clinical psychopathology (trans: Hamilton MW). Grune and Stratton, New York
Shoemaker S (1968) Self-reference and self-awareness. J Philos 65:556–570
Sokolowski R (2000) Introduction to phenomenology. Cambridge University Press, Cambridge
Stephan KE, Friston KJ, Frith CD (2009) Dysconnection in schizophrenia: from abnormal synaptic plasticity to failures of self-monitoring. Schizophr Bull 35(3):509–527
Taylor JG (2000) A control model for attention and consciousness. Soc Neurosci Abstr 26, 2231#839.3
Taylor JG (2007) CODAM: a model of attention leading to the creation of consciousness. Scholarpedia 2(11):1508
Taylor JG (2010) The I's eye view of consciousness. J Conscious Stud 17(1–2):95–117

Taylor JG, Cleeremans A, Freeman WJ (eds) (2007) Brain and consciousness. Neural Netw. 20(9):929–1060
Thompson PM, Vidal C, Gledd JN, Gochman P, Blumenthal J, Nicolson R, Toga AW, Rapoport JL (2001) Mapping adolescent brain change reveals dynamic wave of accelerated gray matter loss in very early onset schizophrenia. Proc Natl Acad Sci U S A 98(20):11650–11655
Uhlhaas PJ, Mishara AL (2007) Perceptual anomalies in schizophrenia: integrating phenomenology and cognitive neuroscience. Schizophr Bull 33(1):142–156
Velligan DI, Kern RS, Gold JM (2006) Cognitive rehabilitation for schizophrenia and the putative role of motivation and expectancies. Schizophr Bull 32(3):474–485
Wynn JK, Breitmeyer B, Nuechterlein KH, Green MF (2006) Exploring the short term visual store in schizophrenia using the attentional blink. J Psychiatr Res 40:599–605
Zahavi D (2005) Subjectivity and selfhood. MIT Press, Cambridge, MA

Chapter 11
The Escape from Life Through Meditation

11.1 The Pure Consciousness Experience

Part of the trouble in trying to understand myself, experienced when I was younger and now by the man I have become is that the soul, considered by many as the centre of the self, is still as vague a concept now as when it was considered in early religions or in Classical Greece or Ancient Egypt. Whatever the mechanism at work to create the soul, its discovery will be the soul's own ultimate coda. Any such proposed coda will be highly controversial. It will, however, be of the utmost importance, for it will then allow the real work of humanity to be done: to face up to this coda for the soul and move towards a true unified humanity. We will then be able to go out into the Universe with the ultimate knowledge about ourselves.

What is such inner experience? This question leads us back to our original roots in religion. There are accounts in religious literature of 'religious experiences' of saints and mystics over the past three millennia. These accounts are similar across many religions – Buddhism, Hinduism, Christianity, Judaism, Islam. They describe a state claimed to be of great value, attainable only by considerable effort, according to past and present practitioners. The accounts are very similar. They talk about what is now called 'Pure Consciousness'.

The best way to describe Pure Consciousness is to imagine your self suspended in a sensory deprivation tank of water. Your eyes are masked, ears plugged and with no outside stimulation. What you will be left with are your bodily sensations, notably from your skin but also internal organs regulating breathing such as the heart and lungs and, of course, your memory and imagination providing you with a stream of thoughts, possibly provoking fear because of the alien environment in which you find yourself. For Pure Consciousness to be achieved, sensory deprivation is taken a few steps further. Step one is to lose sensations of one's physical body altogether and external messages from the skin cease totally. Step two is a drifting away of memory and imagination. Step three, the mind empties itself of all thought and content and becomes a mental state of 'nothingness'. One is in a state of stillness with no physical or mental activity, in the void of oneself. It is the state of Pure

J.G. Taylor, *Solving the Mind-Body Problem by the CODAM Neural Model of Consciousness?*, Springer Series in Cognitive and Neural Systems 9,
DOI 10.1007/978-94-007-7645-6_11, © Springer Science+Business Media Dordrecht 2013

Consciousness that possibly envelopes us in the centre of our being. It is finally our inner Black Hole. No content; it is indeed the nothingness claimed to exist by Jean-Paul Sartre.

This remarkable new state of Pure Consciousness must be one route into the soul, if not the soul itself, if it was to be anything. Was pure consciousness at the basis of religious experience? Certainly, it gave people a sense of the presence of God, according to their accounts. For those with no faith it was still an enormously valuable experience.

But how could it be trapped and an explanation found for it? A partial trapping had already begun by various groups of brain researchers. They had started to look at the brains of those undergoing a state of Pure Consciousness. They found a remarkable common set of brain activities across all such adepts. Subjects all had brain activities in the prefrontal lobe (just behind the forehead, acting as the social controller) and in the parietal lobe (in the brain just above the ears, controlling how motor responses are made to visual stimuli and how attention is kept in focus) as well as reduced activity across sensory parts of the brain at the back and nearby sides of the head. The prefrontal lobe activity could correspond to some goal or objective being activated, such as meditation that cuts out sensory inflow, the parietal activity representing the use of that objective to redirect attention in on itself.

The remarkable feature of Pure Consciousness is that it corresponds to the complete absence of any content of consciousness. But the subject is still conscious. That is why the state has been termed 'consciousness attending to itself'. It could even be called 'Pure Attention'.

So this, I thought, must indeed be the soul or some part of it. But it is such a big step to take, to suddenly find such strong evidence about the soul that we need to consider some powerful support that exists in the writings of mystics across all the world's religions, and written over the centuries. The Pure Conscious State is equated with the 'place where God dwells', although its nature is claimed to be impossible to describe, so identifying it with the nothingness of Pure Consciousness. We must therefore consider the proposal that it is in the centre of the soul that God is to be discovered, through rigorous prayer and meditation, so ultimately through this experience of Pure Consciousness.

11.2 The Common Experience of Mystics: Meeting God

In Christianity, the great mystic Teresa of Avila wrote

> ...this secret union takes place in the deepest centres of the soul, which must be where God himself dwells, and I do not think there is any need of a door by which to enter it.... This instantaneous communication of God to the soul is so great a secret and so sublime a favour, and such delight is felt by the soul, that I do not know with what to compare it, beyond saying that the Lord is pleased to manifest to the soul at that moment the glory that is in Heaven, in a sublime manner than is possible through any vision or spiritual consolation.

Another great Christian mystic, John of the Cross, stated about the union of the soul with God

> In thus allowing God to work in it, the soul... is at once illumined and transformed in God....

Brother Lawrence, another Christian mystic, writes yet again of this mystical union

> When I apply myself to prayer, I feel all my spirit and all my soul lift itself up and without any care or effort of mine, and it continues as it were suspended and firmly fixed in God, as in its centre and place of rest.

These experiences of God are summed up very succinctly by Evagrios the Solitary, who wrote:

> If, then, you wish to behold and commune with Him who is beyond sense-perception and beyond concept, you must free yourself from every impassioned thought.

So in order to commune with God you must annihilate all sensations and thoughts: so what else is the resulting state of mind but that of Pure Consciousness? No content, nothing but experience of 'oneself' directing this totally inward look into one's own (now content-free) mind.

Similar forms of union with God through meditation on the soul are also to be seen in other religions, such as from Abhinavagupta in Hindu literature:

> Once consciousness has been successfully doubled back on itself, this power begins to operate and leads {the aspirant} toward the Ultimate...(and) who is able to unite with the Goddess.

In the Sufi Muslim tradition, Mubhyi-d-Din-ibn 'Arabi similarly writes, in his 'Wisdom of the Prophets':

> Supreme Union is the mutual interpenetration of Divinity and man

Yet all through these writings is the theme of 'annihilation' or 'stillness' as a crucial part of this union with God, as it is a feature in Pure Consciousness. To quote again from Mubhyi-d-Din-ibn 'Arabi, in this state:

> God is mysteriously present in man and man is obliterated by God

It is the process of obliteration that we can see as the process of reducing the mind to the nothingness' of Jean-Paul Sartre.

In The Sufi Path of Love: The Spiritual Teachings of Rumi, he writes

> Union with God is self-annihilation...

Again he writes more fully

> Recognize that your imagination and your thinking and your sense perceptions are reed canes that children cut and pretend are horsies. Deny your desires and wilfulness, and a real mount may appear under you.

In all then, there is claimed to be something deeper inside each of us, only reached by strong enough repression of the outside world and also of oneself as an object. This is beautifully expressed in the writings of The Hua Hu Ching of Taoism:

> The ego is entranced by .... names and ideas... However names and concepts only block your perception of the Great Oneness. Therefore it is wise to ignore them. Those who live inside their egos are continually bewildered.

This theme of absence of awareness in the presence of God is also very strong across many writings of Christian mystics. Angela of Foligno wrote:

> When I am in that darkness I do not remember anything about anything human, or the God-man, or anything which has form. Nevertheless I see all and I see nothing. As what I have spoken of withdraws and stays with me, I see the God-man

Similarly John of the Cross wrote

> ... wherein the soul becomes naught else than an altar whereon God is adored in praise and love and God alone is on it

In Orthodox Christianity, the Abbot Vasilius recounted his experience:

> The soul can attain to the secrecy which is in God, where the mystery of unity beyond understanding and speech is celebrated, and only when it has gone not only beyond the categories of vice and ignorance....

Another Christian mystic, John Climacus stated

> Come to union with the most blessed stillness and I will teach you the workings and the behaviour of the spiritual powers

We come back to Teresa of Avila for a conclusion, from what is only a small fraction of the relevant writings from the mystics of the past millennia:

> ... in this temple of God, in this Mansion of His, he and the Soul alone have fruition of each other in the deepest silence.....The faculties, I think, are not lost here, it is merely that they do not work but seem dazed

There are many more religious writings in a similar vein. They all point to the absence of mind and connection to the world around us. So the mystics of the world's religions were all experiencing this claimed union with God at the centre of their soul when they entered into the state of Pure Consciousness. They did not know of what they were experiencing, except they knew they were experiencing something, even as an absence of all content, That is the state of so-called 'annihilation', of 'stillness', or as is called in Eastern traditions, Samadhi or Nirvana. It appears to be universally experienced by many mystics of the past.

11.3 The Ground Truth of Religion

I cannot leave the claims of these mystics and meditators of the past without considering their validity. Let us assume that these dedicated people were able to meditate or pray to get them into this remarkable content-free state. They had no sensations of the outside world or of their own body. This is a remarkable (and stress-free) state. But did they actually meet God when in the state of pure consciousness? Or were they interpreting the total lack of experience in this state by bringing a modicum of colour to it from their own religious backgrounds and experiences? Each of them brings an appropriate interpretation according to their indoctrination and belief system. Why not?

Consider how, in each religious persuasion, the interpretations given by the mystics are imbued with the nature of their religions: Jesus Christ or the God-man for the Christian, Allah in Islam, the Great Oneness in Taoism, and so on. It seems natural therefore to suggest that these 'biased' interpretations are naturally superimposed on an underlying core experience, common to mystics in all religions. That core is 'stillness' or 'absence'. It can easily be filled by the mystics through their prior indoctrination in their religion, as from early childhood. This world-view will naturally colour experience, especially if it is of nothing at all. So the appropriate religion fills in the dark empty spaces of pure consciousness, imbuing it with the colour of the particular religion in each case. And such imbuing gives the mystic support for his or her religious view, reinforcing it every time they enter the pure consciousness state. It is their rock-hard 'ground truth'.

What about the rest of the world's religious community – those many others of us who are not mystics, nor have ever tried to be? How do they get their ground truth? Not through the very hard work as performed by the mystics to meditate into pure consciousness and so commune with their God! Are the rest of us just hangers-on to the tail-coats of these trail-blazers? Are the religious beliefs of non-mystics – supported by the outpourings from the mystics – being ritualised in the creeds of their faith, so reinforcing the idea that they have an automatic response to a variety of personal and social situations? They have the examples of the mystics of their particular faiths indicating that their God can be reached by a special few. But these few do actually exist, so providing proof that their God also exists. Correspondingly, the God of their mystics espouses similar basic tenets of faith (the existence of the Holy Trinity, reincarnation, etc, for Christians). So these believers, whom I call 'ritualists' are doubly supported in their faith: firstly through the repetition of prayer, mantra and ceremony to reinforce their belief set, and secondly through reported personal ecstatic experiences of their mystic leaders.

The conclusion of my analysis is that the real ground truth of religion comes from the mystics; all the others follow the rituals devised to enhance their beliefs in their religion but gives them no direct experience of God or even any claim to it. But there is still the ground truth of the mystics' experiences, with commonality across all religions, and with local variations thrown in where needed. No wonder I could not participate at first hand in the nature of God when at my communion classes as a young man living in the wilds of Essex. If the vicar had trained me in meditating to stillness I might then have experienced the God-like items he talked about. But he did not do so, and I found out I could not be a ritualist.

It could be claimed that my approach through pure consciousness to the heart of religious belief leaves out completely the records of the Great Books: the Bible, the Koran etc. It is the case that these books provide a framework of beliefs around which one can weave one's own special beliefs with the relevant Book providing back-up. But these Books were written thousands of years ago. It is the mystics who have the power to bring the relevant set of beliefs to life. Through their probing of their experience of Pure Consciousness they can bring to their circle of admirers the reality of their religious experience.

If pure consciousness gave direct access to God then it would be of great value to all. But as noted earlier, it is unlikely that the state of pure consciousness can give such an unadulterated access, since it was noted that the God met by the mystics in this state varies from religion to religion according to the religion of the mystic undergoing the experience in a perfectly explicable manner. It could be argued that the God that is being accessed by the mystics is the same for them all, and they are only putting a different spin on their experience. It can even be argued, from many of the statements of the mystics, that they do not experience their God in the midst of their stillness, but only interpret such an experience by the external cloth of their Book put on it by themselves.

Does pure consciousness, then, finally give an experiential proof of the existence of this common God? We cannot answer that important question easily. To do so requires further work of looking at the brain itself, as to how there could be any interaction with an entity called God. The numerous questions that then arise in trying to answer this important question are to do with the nature of such an interaction. Can God's finger poke into the brain ever so gently to bring on this mystic state? Where could that 'poking' be seen in the brain –which we must assume occurs for the mystic to have the experience in the first place. If it does not affect their brain in any way then they would know nothing about it. This would be contrary to the reports I recounted above from so many mystics.

Can people develop special ways of inducing this interaction (as in the case of mystics entering pure consciousness)? How can we interpret God's model of humans through his supposed interaction with our brains? These and many other questions are open to be considered through this amazing phenomenon. But they all involve sensitive and difficult questions associated with the way the brain works in these and related situations which we cannot address presently.

Of course if we are truly made in God's image then finding out about the brain sites and their functionality to achieve Pure Consciousness might truly be regarded as attempting to discern the 'Mind of God'. It has been claimed by some scientists that we might observe this 'Mind' at work in the Big Bang. Why not also in the state of Pure Consciousness, as claimed by the religious mystics I quoted earlier? Indeed such an identity would be more likely than trying to see any intelligent and conscious God-like entity at the beginning of the Big Bang. As we emphasized in Chap. 1, in the beginning the Big Bang was totally inhuman. It would have no relation to any God, especially one that was claimed to have been creating humans in his own image. The power available at the very beginning of the Big Bang would have been so enormous as to blast any human-like creature out of existence. A similar fate would occur to any of us if, for example, we stood in the beam of the Large Hardon Collider (LHC) in Geneva (perish the thought): instant annihilation of anyone so absolutely stupid. No contact at all with the Mind of God. The intrepid adventurer trying to meet God in that way would just be vaporised into the new matter to be observed at the LHC in due course.

I had related the soul to Pure Consciousness, by evidence from mystics through the ages and the religions of the world. It was there, if anywhere, that God was to be met most closely. The soul was thus capable, I thought, of at last being closely related to purely neural activity. Such guidance in the brain that could occur from the finger of God could possibly be detected by building a very detailed model of the dynamics of the pure conscious state. If that model requires unexpected or external inputs (as from God's finger) to make it work then that would be astounding. Instead of trying to weigh the soul of the dying tuberculosis patient (MacDougall 1907) we are now trying to see in more detail how the soul could have support from the brain. Of course we are a long way from such a position – suitable brain models are now being conjectured, but it will be decades if not centuries ahead before we can try to probe for the 'Finger of God' needed to explain any unexpected and unexplained changes (which will also have to be observed experimentally). But at least it gives us a direction in which to probe these deep questions. It may help give some sense of direction to the search for the soul to many who have been following it for many years.

But we are getting ahead of ourselves. The brain has to be able to support such a unique state as pure consciousness. We can begin to probe that state from the outside. But then we need to know what signature would make that state detectable. We mentioned it has been scientifically observed, but no hint of its relation to the soul been hinted at. Moreover in order to properly understand pure consciousness, we need a working model of it. How could such a remarkable and subtle state be supported by suitable functionality of the brain? To discover more on this, we must widen our search for the soul. We have to develop an understanding of the whole range of experiences that the brain can support – thinking, reasoning, feeling strong emotions, sleeping and dreaming, under drugs, in states of mental ill health such as schizophrenia, and so on and on. Most specifically how does the brain achieve this wide variety of experiences? And even more specifically how does it achieve our normal consciousness for the seven billion of us here on earth? Even more remarkable how did that inner self evolve. Lower animals such as the worm do not seem to possess such an experience. Higher up the evolutionary scale perhaps the soul creeps in. Is such a soul, at a lower level, possessed by your cat or dog? That requires exploration since it may give new hints on how the soul itself evolved. Indeed new windows on the mind!

11.4 Explaining Pure Consciousness by CODAM

If the CODAM model of consciousness is at the basis of human consciousness, then it should be able to explain the mechanism underlying the pure consciousness state. That sounds like sacrilege after what I have written about the experiences of mystics claiming to be in communion with their God. However there has to be some mechanism by which this communication can take place, with its associated

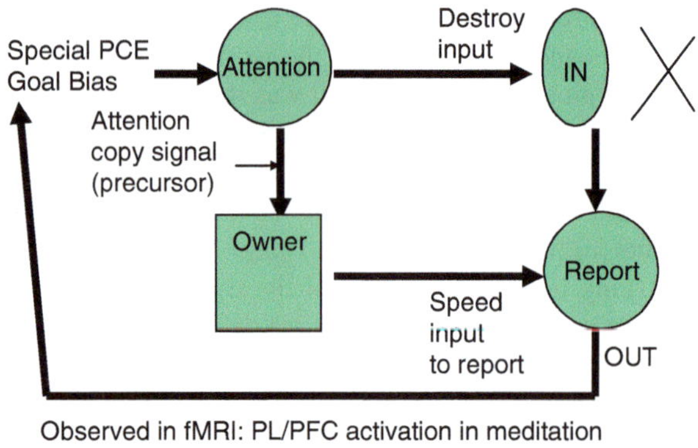

Fig. 11.1 Modifications of the CODAM architecture to obtain the PCE

detailed brain basis. It seems that CODAM may be a good place from which to start to determine if the Pure Conscious State is one of those supported by some variety of CODAM.

The main feature of the Pure Conscious State is that it supposedly has no content. Thus the inhibitory feedback achieved by the corollary discharge signal must be trained to extend to ALL input, either of environmental or internal sensory form. Such an all-inhibiting signal will be expected to take a considerable time to train to be effective. This is exactly as observed in the process of meditation, either in religions such as Buddhism in the East, or in the Western world in a more secular framework. The learning process will be expected to take years to complete, which is exactly as observed.

One such approach (along which I went for a while) is that of 'one-pointed attention'. Here one learns to attend to a single point, by continued meditation on that chosen point (whatever it is- say a cup or a picture or a flower). Only when one has learnt to attend to that single point does one then attempt to annihilate attention to that single chosen object. It is this last step which is crucial, but difficult. But this is only one way of approaching the Pure Conscious Experience.

Having learnt to re-route the attention corollary discharge signal to remove all input it is necessary to learn the goal to achieve that over a long period, as well as holding to that goal. The meditator has no specific rules as to how this special goal is to be achieved, perhaps explaining the difficulty of the overall process. But it is clear that such a goal can exist, and would give the meditator exactly the experience of Pure Consciousness. A possible circuit for this is shown in Fig. 11.1

In the figure we have shown how the attention feedback signal of CODAM (see Fig. 6.4) has been trained (through meditation) to inhibit all inputs. At the same time a signal from the working memory site (or from the attention movement signal generator) has learnt to create a suitable goal signal to cause the input inhibition, and so attain the PCE in a stable manner.

We conclude that CODAM can form a useful platform to allow for a range of conscious experiences, not just of the Pure form but also of the even more advanced one in which both an inner stillness concomitant with knowledge of the outer world

(Taylor 2002) being possible. Even an explanation of the Oceanic Consciousness State (Werman 1986), in which a person seems to inhabit all the components of the world he views, as well as his inner self, could be explained by similar means. In this Oceanic state he would be fusing his sensory states of the external world with his own personal experiences, in an ineradicable manner.

11.5 What Is the Value of Pure Consciousness?

This is a question asked by the Buddhist Monk when he went into a Monastery where in one room he found a number of monks sitting silently in their ‘stillness’ state. He kicked one of them, saying they were worth no more than a stone sitting there. One can ask today of the meditators who have gained this level of internal peace: how many of them have obtained Nobel Prizes in Science? How many have obtained prizes for great works of Art? My answer would be: None. Not that the meditators are meditating towards such ends. But then it would appear they are meditating to get away from life, not to enhance it. I think this discussion has shown that the title of the chapter is apposite.

It might be asked by the reader, however, that if Pure Consciousness has little value then why has space and time been spent on it at this juncture or even at all in this book. The answer to this position of denigrating Pure Consciousness is that it is clearly an experience of some import to the people who attain it (‘nirvana’ or paradise). Besides that, it is an extended experience of ‘stillness’, which can be regarded as a direct experience of the ‘nothingness’, claimed by the Western philosophers of mind mentioned in Chap. 7 as being the core feature of the inner self. So it would seem that those who experience Pure Consciousness are a living instance of this important feature. They are directly experiencing their ‘I’.

There are several difficulties in using this result as proof of the existence of ‘I’.

1. The ‘I’ of Pure Consciousness may well be heavily distorted version of the day-to-day ‘I’, since the former is only achieved through the considerable hard work of meditation over numbers of years.
2. Since the Pure Conscious experience apparently consists of absence of experience, then it is difficult to tie down from personal report that such experience (if there is any) in Pure Consciousness is at all similar to that in daily life.
3. The Pure Conscious experience has much religious bias, and is quite different from what we would expect the ‘I’ to be for the average person.

We thus still need to take seriously the ‘problem of I’ as raised in Chap. 8. Yet Pure Consciousness is a further clue in support of the existence of ‘I’ and is also an intriguing case of an altered form of consciousness helpful in further exploration of CODAM.

References

MacDougall D (1907) Hypothesis concerning soul substance together with experimental evidence of the existence of such substance. J Am Soc Psych Res 1(5):237–275
Taylor JG (2002) Paying attention to consciousness. TICS 6(5):206–210
Werman DS (1986) On the nature of the oceanic experience. J Am Psychoanal Assoc 34(1):123–139

Chapter 12
The Evolution of Human Consciousness

12.1 Introduction

Determination of the details of how consciousness evolved in living systems is still to be achieved. In this chapter we take an approach to this problem through the evolution of the mechanism of attention across the animal kingdom. This is natural since one must attend to a stimulus in normal viewing conditions in order to be conscious of it. The present understanding of attention in primates, using single cell and fMRI results is shown to lead to a control approach which we suggest evolves through four stages to an extended form of ballistic attention control; a similar sequential approach was already discussed in Chap. 5, although will take a slightly different form here due to the approach trying to keep to an evolutionary path rather than purely a functional one. The final step in either path is to be finally extended for humans by addition, we propose (on the basis of the CODAM architecture) of a corollary discharge signal of that for the movement of the focus of attention, as developed in Chap. 6. It is suggested that this is the final (fifth) stage in the evolution of attention, with consciousness as a concomitant of the corollary discharge signal itself.

We appeal to comparative neuroanatomical features as well as paleoneurological results in order to begin to fill in some of the details of such conjectures. We explore how such stages of attention may occur in the animal kingdom, leading to at least four levels of attention control being present in primates. Evidence from brain imaging indicates that an inner self may be absent in the macaque, but is present in humans, as discussed in Chap. 9.

In order to fill in some of the intermediate steps in this progression, paleoneurological evidence from changes in cerebellar size is taken to help explain the Upper Paleolithic Revolution, with the sudden development of much greater sophistication in tool use, personal adornment, etc.

J.G. Taylor, *Solving the Mind-Body Problem by the CODAM Neural Model of Consciousness?*, Springer Series in Cognitive and Neural Systems 9,
DOI 10.1007/978-94-007-7645-6_12, © Springer Science+Business Media Dordrecht 2013

12.2 The Evolution of Mind

There have been various attempts recently to understand how the mind might have evolved in Darwinian terms (Gangestad and Simpson 2007; Geary 2004; Lewis-Williams 2002; Jaynes 1976; Buller 2009). More detailed approaches have used the methods of both paleoneurological and comparative neuroanatomic information taken from the brains of living vertebrates (Jerison 2007; Nieuwenhuys 1994) as well as discussing the possibility of consciousness in presently living animals (Edelman and Seth 2009; Butler 2008; Allen 2009). These are all important steps to take towards completing a detailed account of evolution already well developed through a painstaking study, taking place over many years, of fossilised remains of the skulls of our numerous ancestors. It is the important missing link of the evolution of the mind of living beings that we consider here: how did our consciousness evolve as part of the evolution of the brain?

It is true that there are difficulties in such a pursuit through paleoneurology due to the non-persistent nature of brain tissue, although general changes of skull shape allow for hints to be gleaned as to how the brains of animals have changed over geological time. Overall skull size has allowed much to be conjectured about the alterations in mental power correlated to such changes (Jerison 2006). For example the evolution of brain size has been analysed in detail for fossils from periods over the last several hundred thousand years in their relevance to understanding the development of cave-painting and related behaviours (Lewis-Williams 2002). Understanding has thereby been reached as to how increased brain size has related to augmented social pressures; such pressures have been considered to cause the evolution of new brain-based functionalities to handle the increasingly competitive environment both from weather changes to increased social competitiveness and other sources (Lewis-Williams 2002; Gangestad and Simpson 2007). In this chapter we wish to develop a related analysis that takes account of brain functionality in a different manner, concentrating on specific brain functionalities which could have led to the evolution of the human mind and in particular ultimately to human consciousness.

The above references do not consider the mind in its fullest sense – that supporting conscious experience. That there are such and such extensions of brain tissue to support the enlargement of such and such behavior patterns are to be expected, and can be detected by correlations between the brain expansions (from brain endocasts) and artefacts detected in the homelands of the relevant species. This approach has given much information. But the fundamental question was that already asked above in our title, but not answered. As Darwin (Darwin 1871, 2004) wrote in 'The Descent of man':

> Nevertheless the difference in mind between man and the higher animals, great as it is, certainly is one of degree and not of kind. We have seen that the senses and intuitions, the various emotions and faculties, such as love, memory, attention, curiosity, imitation, reason, etc., of which man boasts, may be found in an incipient, or even sometimes in a well-developed condition, in the lower animals. They are also capable of some inherited improvement, as we see in the domestic dog compared with the wolf or jackal. If it could

> be proved that certain higher mental powers, such as the formation of general concepts, self-consciousness, etc., were absolutely peculiar to man, which seems extremely doubtful, it is not improbable that these qualities are merely the incidental results of other highly-advanced intellectual faculties; and these again mainly the result of the continued use of a perfect language. At what age does the new-born infant possess the power of abstraction, or become self-conscious, and reflect on its own existence? We cannot answer that question yet, nor the question in regard to the ascending organic scale. The half-art, half-instinct of language still bears the stamp of its gradual evolution. The ennobling belief in God is not universal with man; and the belief in spiritual agencies naturally follows from other mental powers. The moral sense perhaps affords the best and highest distinction between man and the lower animals; but I need say nothing on this head, as I have so lately endeavoured to show that the social instincts—the prime principle of man's moral constitution—with the aid of active intellectual powers and the effects of habit, naturally lead to the golden rule, 'As ye would that men should do to you, do ye to them likewise'; and this lies at the foundation of morality.

So from Darwin we conclude that consciousness did evolve along with other brain-based powers. In this chapter we attempt to develop a detailed modus operandi for such evolution and present support for that from available fossil data.

The most important problem for a freshly-evolved mobile animal is that of surviving in a hostile but complex environment. There will be many stimuli in its surroundings that it will have to take into account. This confusion could be handled by attempting to notice all of these stimuli at once. But such an all-seeing approach seems inefficient in comparison to evolving a facility allowing rapid filtering of the environment so that only one or very few stimuli need to be considered in detail at once – the most dangerous, for example. The animal can thereby handle novelty in small doses.

Such a filtering facility is that of attention, divided into a top-down version able to direct attention according to some goal ('watch out for the bogey-man') and a bottom-up one ('hey – what's that looming object suddenly appearing above me'). These two components – termed endogenous and exogenous attention respectively – are now an important area of behavioural and physiological study, especially in humans. How attention has evolved both over geological time and also in a comparative manner is crucial to be teased out in order to appreciate the manner in which external objects in normal viewing situations by humans only become conscious if they are attended to. We have already addressed this question in some detail in Chap. 3, but then only for humans. Here we assume that consciousness, at whatever stage it emerged through evolution, was already embedded in its attentional cover.

Consciousness, as possessed by humans, requires attention to be alerted for normal viewing situations. If we wish to include in the mind the beautiful but subtle conscious experience possessed by humans (and possibly some lower animals) we must therefore be prepared to study the evolution of attention as a prerequisite to understanding the evolution of the conscious mind. This is a completely different emphasis on the evolution of mind as compared to previous accounts, such as those referred to above, which depend heavily on behavioural evidence. It is an approach that could give new insights into the complete evolutionary story of the mind. Of course the evolution of attention has itself very likely gone hand-in-hand with that of ever more effective behaviours in response to a complex environment (such as with the growing powers of internal models in the evolving brain allowing greater reasoning powers). But we are considering here the underlying brain modifications that were needed to achieve this evolution, especially that of the faculty of attention itself.

12.3 The Evolution of Attention

We make here the main conjecture: consciousness arose from an increasingly sophisticated and evolving version of attention, allowing for greater speed in attention deployment as well as an added ability for self-correction and self-knowledge. Such added features would undoubtedly provide an increased evolutionary advantage to any animal possessing them. Through probing and properly understanding the detailed evolutionary development of attention we propose that the evolutionary journey to consciousness was accomplished. It is a journey we try to duplicate in this chapter.

It involves also noting that we would expect a similar evolutionary increase of representations of stimulus activities, as seems to be the case in the visual cortices and temporal lobes in animals up to humans, and is also seen in the spatial maps in the dorsal visual route in a similar evolutionary progression up to humans.

We have already (in Chaps. 5 and 6) developed a sequence of stages of attention in order to be able to extend them from the extended ballistic attention control model described in Chap. 5, to CODAM explored in Chap. 6. Here we will consider such a sequential analysis but now from a slightly different viewpoint: that of how evolution may have driven the development, and especially specific architecture of the various modules.

In order for such competitive architectures as evolved to be adaptively viable for representations of input stimuli, the competition between input stimuli representations would most effectively be run on a separate neural module than that containing the initial input coding of the stimuli (the latter being such as the retina or lateral geniculate of the thalamus in vision). In this manner, various stimuli can be input to the brain, with the winner of a competition on the saliency map being obtained without necessary loss of other input stimuli. These latter would then be more rapidly available than if they had, all but the winner, been lost in the competitive process. This competition could be by some form of saliency map, as proposed in the original attention model of (Koch and Ullmann 1985).

We regard this architecture as the evolutionarily most primitive form of attention, where the winner on the saliency map can be directly employed by the animal to plan actions associated with the attended stimuli. The corresponding architecture is shown in Fig. 12.1. There is a possibility that such a salience map is detectable at a low level in the visual hierarchy in the human brain (Li and Dayan 2006), such as in V1, V2 or V4 (Fragopanagos et al. 2009).

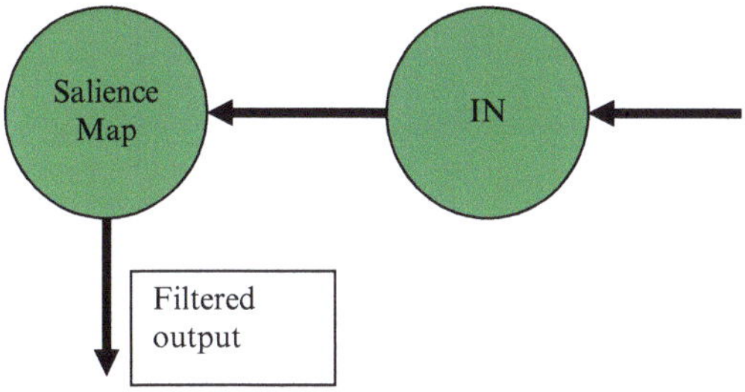

Fig. 12.1 Basic saliency map control. The output is taken from the saliency map, which could be at the level of V1 in the visual cortex (Li and Dayan 2006). This architecture is taken as stage 1 of the evolution of attention

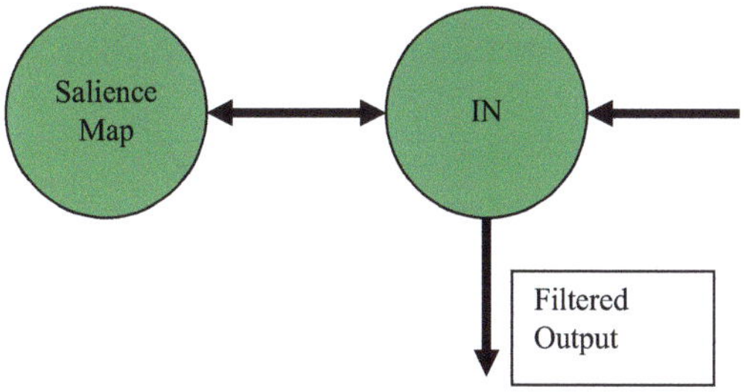

Fig. 12.2 The extended saliency map attention controller. The output here is taken from lower level input feature analysers, after a competition in the higher order saliency map has feedback support for the most salient input in the IN module. This closely follows the earlier suggestion of (Koch and Ullman 1985). This architecture is taken as stage 2 of the evolution or attention

The next stage in the evolution of attention would be expected to allow for the possibility that the saliency map winner feeds back its activity to cause more efficient processing of the activity representing the attended stimulus. The resulting stage 2 would thereby use more neural connectivity, and hence a larger brain, as well as possibly allowing for a higher level saliency map. A minimal architecture for this is shown in Fig. 12.2. Such an extension of the architecture beyond that of Fig. 12.1 would also be able to make for more independence of the saliency map. As such this would allow for increasingly stronger biases entering the saliency map, so allowing for more sensitivity to the environment as well as for the development and employment of internal goals.

The next stage 3, of the evolution of attention, has a simple architecture for it shown in Fig. 12.3. Its occurrence as a separate evolutionary stage is supported by its presence as the second developmental stage in infant attention development (see Sect. 12.5). It involves a top-down goal that is used to bias an attention signal to filter the lower-level target representation.

The succeeding evolutionary stage would be that allowing the animal to store its attended stimulus activity, suitably coded, on a separate module for further use. This clearly would involve more brain capacity, as well as recurrent connectivity

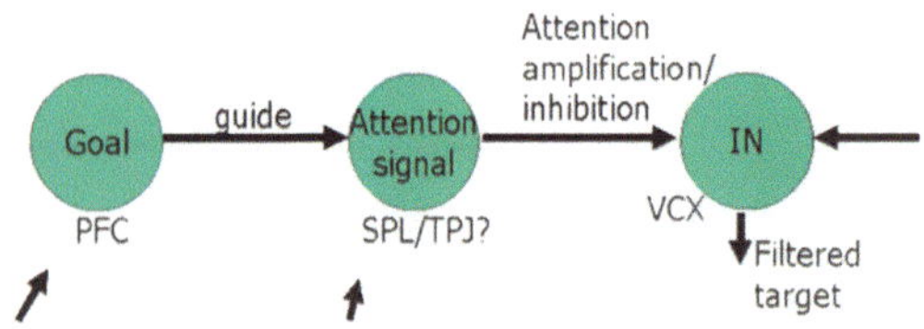

Fig. 12.3 The ballistic control model of attention. *PFC* prefrontal cortex, *SPL* superior parietal lobe, *TPJ* tempero-parietal junction, *VCX* visual cortex. This is based partly on the 'biased competition' model of (Desimone and Duncan 1995). This architecture is taken as stage 3 of the evolution of attention control

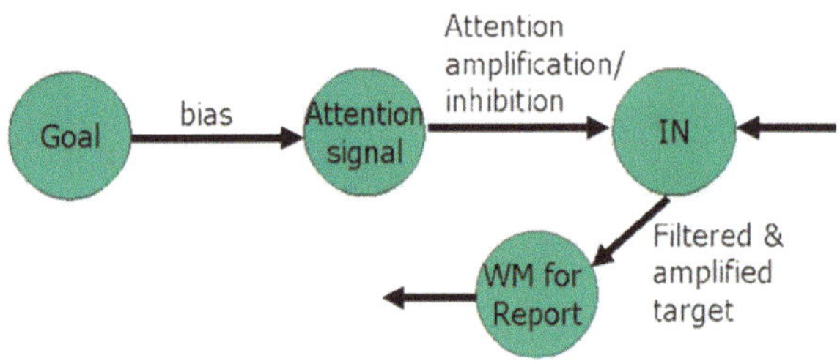

Fig. 12.4 The extended ballistic attention control model. There is activity on the extra working memory buffer (beyond the modules in Fig. 12.3) allowing for further use of the representation of the attended stimulus, for reasoning, thinking and so on, by the subject. WM, working memory. This architecture is an extension of the model of (Desimone and Duncan 1995) by addition of the WM module as a short-term memory. This architecture is suggested as that of stage 4 of the evolution of attention control

to allow for longer decay of the neural activity winning the competition. Moreover the held activity is expected to be filtered by attention amplification of the stimulus representation so that only a suitably active representation would gain access to the longer-lifetime module, usually termed a short-term memory (STM) or working memory buffer and well supported by observations in human brains (Xu and Chun 2006). Thus we come to the extended ballistic attention control system of stage 4, which is that of the biased competition model of (Desimone and Duncan 1995) augmented by addition of a short-term or more complete working memory. In this extended model higher-level goals are chosen from among the stimulus representations set up (by learning) in lower cortices, but now used to bias the attention feedback signal in a typical search task, for example. A possible neural architecture for this is shown in Fig. 12.4, with an extension beyond the original model involving a pre-frontally-based working memory module for goal-holding.

The evolution of a module for holding attended input in an STM is to be expected to be somewhat concomitant with the use of these 'goals' in the biased competition model. In other words stage 3 is expected to be roughly evolving at the same time as stage 4. The earlier possibility (stage 3 before stage 4) would be so if the prefrontal Goal-held activities of attended stimuli were to be used in planning other than the direction of attention; the parallel development of stages 3 and 4 arises if the STM attended activity were used immediately for biasing further attention processing (thus acting as goals). However the process of further planning seems more complex than simple use of continued activity for guiding attended movement, so we conjecture that stage 3 evolved before stage 4.

This possibility is supported, as noted above, by the observation in infants of a developmental second stage of attention. In this stage, putative prefrontal goal activations are not yet observable in the infant brain, although short-term memory in the parietal lobes has been detected.

12.4 The Final Evolutionary Step: Consciousness

In the process of developing a viable evolutionary set of architectures for attention across time we have arrived at the extended ballistic attention control model of Figs. 12.3 and 12.4, which we conjecture are the beginnings of conscious experience created in an animal possessing such architecture. The conscious experience would be generated by the continued activity on the short-term memory site, carrying an attended stimulus representation which had attained a suitable level of activity by attention amplification. This activity could thus act as part of a global workspace type of activation (Baars 2002, 2005), being available to other modules for further information processing. Such an approach has been considered to be a basis for consciousness of the content as carried by the relevant representation.

We immediately come up against the hard problem of consciousness (Chalmers 1996), in terms of the question: for whom is the corresponding activity conscious? This was raised by (Nagel 1974) in his well-known question: "what is it like to be a bat?" and underlined by the presence of the explanatory gap in (Levine 1983). The basic problem is that there appears to be no owner of the relevant conscious experience. There would therefore be no conscious 'experience' at all. Only by the neural creation of an owner of the 'conscious' content of attended stimuli could consciousness be claimed as present. We need to explore possible neural architectures which would lead to such ownership and hence allow us to begin to approach the culmination of the evolutionary process that, we assume, led to human consciousness.

We have already faced up to these problems in Chap. 6. We discussed there how the only such approach based on attention is that of the CODAM model (Taylor 2000, 2002a, b, 2007, 2010a, b, c, d), as was presented in Chap. 6. We consider here, then, using attention more directly but extending it to allow for use of the attention copy signal, as in the CODAM model. It is to be expected that a copy of such an important signal would be sent to other higher-level areas. Such a signal could speed up access of the activity of the attended stimulus to its short-term memory site, as well as prevent distracters from causing errors in response. It would also alert other neural areas to prepare for the arrival of conscious activity for further processing. Such an architecture is shown in Fig. 12.5 (identical to that of figure 6.3 in Chap. 6). As noted earlier, simulation of the

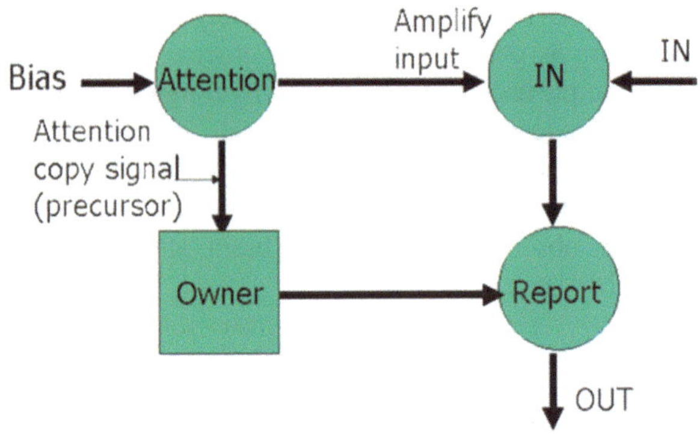

Fig. 12.5 The CODAM (COrollary Discharge of Attention Movement) model. The added module beyond the architecture of Fig. 12.4 is that of the STM for the corollary discharge of the attention movement control signal, termed in Fig. 12.5 the 'owner' module. The purpose of this further module is to provide the corollary discharge signal to be used for error correction and distracter removal on the input signal of the amplified representation of the attended target stimulus, and was introduced initially in Chap. 6. This architecture is considered as the final stage in the evolution of consciousness in humans

CODAM architecture was employed in a successful model of the attentional blink (Fragopanagos et al. 2005) and important components of it, suitably extended, in a simulation of the phenomenon of distracter devaluation (Fragopanagos et al. 2009). These points were discussed more thoroughly in Chap. 6, to which the reader is referred.

12.5 Evidence from Evolution and Comparative Neurology

12.5.1 The General Picture

To summarize the position we have reached in our search to understand the evolution of consciousness, we present in Table 12.1 the possible evolutionary steps we assume attention has taken, as discussed in the previous two sections.

We now need to see what possible evidence exists for the evolutionary stages developed in Table 12.1 from paleoneurology or comparative neuroanatomy There is little evidence in direct fossil terms, but the general correlation of brain processing with brain sizes of ancient animals as they evolved can be taken partly from the capabilities of known living species of a similar form. Such an approach has in particular been discussed recently in (Finlay et al. 2001; Jerison 2006) from a functional point of view.

Table 12.1 Functionality and possible brain sites of the evolving attention circuitry

Stage of attention development & architecture	Necessary components	Possible brain sites
Stage 1: Saliency map filter control of input–output. Details are given in Fig. 12.1	Input analyser; salience map for filtering; attended output of the map as winner of a competition	Brain stem and lower Visual Cortex
Stage 2: Saliency map moved to higher level to send feedback signal to lower Visual Cortical sites. Details are given in Fig. 12.2	Higher level of saliency map and feedback from saliency map to lower Visual Cortex to help competition	+ Further higher Visual Cortex & in temporal or parietal lobe
Stage 3: Attended target activity storage addition as STM. Details are given in Fig. 12.3	+ Site for holding attended target activity, acting as an STM	+ STM in parietal to hold suitably attentionally-amplified target activity for a short period
Stage 4: Extended ballistic control. Details are given in Fig. 12.4	Ballistic control by PFC goal-based longer-term biasing of attended stimulus	+ PFC for goals as WM, to bias the winner of the competition for the focus of attention
Stage 5: Attention copy model (CODAM). Details are given in Fig. 12.5	Storage and use of the attention copy signal for error correction and attention speed-up of access to the visual STM	+ Corollary discharge and STM for it

The specific architecture is taken from attention in vision, although other modalities could be considered in a similar manner

PFC prefrontal cortex, *STM* short-term memory, *WM* working memory, *CODAM* corollary discharge of attention movement

12.5.2 Invertebrates

We first note that there are only two groups of invertebrates with suitably complex brains to make them worth considering: arthropods (insects, arachnids, crustaceans etc) and cephalopods (octopuses and squids, for example). Arthropods have a central brain, whilst cephalopods can have very complex distributed brains, as for the octopus; such brains may be as complex as some vertebrate brains.

The most interesting cephalopod is the octopus. As solitary hunters, octopi are exploratory, attending to any novel object in their visual field. A careful study has been made of various cognitive features of the octopus, with its impressive 500 million nerve cells in total and 40–45 million cells in its central brain (Hochner et al. 2006). Whilst there is some understanding of learning and memory in the octopus (Hochner et al. 2006) its visual attention circuit is unknown. However the animal is clearly able to select a future meal very quickly so expected to possess a very efficient attention system. How this fits into the stages 1–5 outlined in Sect. 12.2 and Table 12.1 needs further comparative data, but from what has been reported in numerous publications about the nature of octopus abilities we expect that they and possibly some cephalopods and high-level arthropods possess attention at least to stage 1 and in some cases up to stages 3 or 4 (as in the case of the octopus).

12.5.3 Vertebrates

Let us now turn to vertebrates. Most vertebrate brains are known to possess about the same brain subdivisions (Northcutt 2002), these being a cerebellum (for fine tuning of actions and sequence learning and control), cerebral hemispheres(controlling sensory and action processing), thalamus (relaying information, motivation and action generation involved in consummatory actions), hypothalamus (controlling sleep/wake cycles, eating and drinking control and hormone release), medulla oblongata and spinal cord (sensory and motor functions), olfactory bulb (specialised for olfactory processing), hippocampus (for long-term memory), and finally the basal ganglia (for action selection and learning, especially from changing from controlled to automatic responses). A number of these centres will clearly be involved in attention in one of several possible modalities, especially as way-stations for relevant information to be transmitted to higher (cortical) attention centres. However, experiments on decerebrate animals indicate that the most important component of attention control is in the cortex.

In terms of Table 12.1, we conjecture that all vertebrates possess attention at stage 1. There are those that go to much higher stages, as we will discuss shortly, where humans are expected to go up to the highest stage. In particular we will consider the mammalian brain due to its efficient control of the animals' behaviour.

12.5.4 Mammalian Attention Control

Mammals can be observed to have attention control systems of a variety of levels in efficiency. This is clearly of great importance either for predators or prey, so attention is expected to have evolved over past periods in a concomitant manner between predators and their prey, following the corresponding evolutionary pressure: an upgrade of attention control in a prey species will require a corresponding update in its predators to be able to catch it, and vice versa.

As concluded by a careful experimental/modelling study (Hilgetag et al. 2001, p1286) "spatial attention in mammals is produced by a wide-ranging, topographically organized, bilateral network of specialized brain regions. This network displays competitive bilateral interactions, essentially mediated, at least in the cat by mutual inhibition of subcortical structures". However the results on the importance of a structure such as the superior colliculus in the brain stem in attention processes in cats is to be expected, since there is a mixture of covert attention and eye movement control involved in the experiments reported by Hilgetag and colleagues (and used in the simulation). We meet here the need to separate the covert attention control system in the mammalian (vertebrate) brain from the motor control of eye movements. These two circuits are known to employ rather similar neural tissue, although there are important differences. One of these is in the crucial use of sub-cortical structures (as in eye movement) compared to cortical networks (as in covert attention movement). However for the latter we should not neglect the numerous cortico-thalamo-cortical loops that will allow both for more efficient as well as faster and broader spread of information across the cortex.

It has been shown experimentally (Sumner et al. 2002) that involuntary attention shifts do not require signals in the direct collicular pathway, so would be expected to arise from some cortical control system. There is also evidence for both basal ganglia and thalamus involvement in attention shifting, but again these are expected as part of the recurrent cortico – striatal – thalamic – cortical loops involved in the cortical dynamics.

From these results we may expect that the majority of mammals have attention control systems at least up to stage 3 or stage 4, with associated working memory systems. We have already raised the question of the difference between levels 3 or 4; we noted these two stages could well be fused, as the evolution of a goal-holding architecture would require a concomitant STM system to allow manipulations to be made (such as refreshing the neural activity or rotating or otherwise transforming it, as in the N-back task in humans). A more detailed paleoneurological study will be needed in order to assess this possibility in the evolutionary past.

We need to turn to primates in order to understand in them the evolution and spread of attention control, and hence of consciousness, before we go back to more general mammalian or vertebrate examples.

12.5.5 Primate Attention Control

From present behavioural studies we would expect that modern primates have attention powers at all stages 1–4. They clearly have goals and their related neural structures. They also have some STM capabilities. We do not know, however, if these functional components both evolved simultaneously, or if one or other of the two components for stages 3 and 4 emerged first and the other second in the evolutionary race.

There has been considerable research into both living primates and the fossil evidence for evolution over the last 100 million years. However a detailed study of brain evolution based on fossil brains is difficult due to the problem, as lucidly expressed in (Weaver 2005, p. 3576) "endocranial markings are difficult to identify", so that "functional inferences based on sulcal patterns are problematic". This implies that it will be difficult to detect changes in the attention network of the brain as it has evolved over the last few million or so years. Yet there were important features of the ratio of cortical to cerebellum volumes discovered by Weaver, to which we will turn shortly.

Looking again at the comparative anatomy of living animals should help in our quest. A recent highly relevant paper is that of (Husain and Nachev 2006), where a careful discussion has been given of the possible homologies between macaque and human parietal lobes. It is suggested there as a serious possibility that the human superior parietal lobe (SPL) has a strong homology to the whole of the monkey posterior cortex (PPC), and that the human inferior parietal lobe and parts of the intra-parietal sulcus (IPS) may have no similar regions in the macaque cortex. As they write (Husain and Nachev 2006, p 33): "There are, in addition, regions of human PPC that seem not to have clear homologues in macaques. Comparative anatomy considerations suggest that the IPL has expanded greatly in humans compared with monkeys, particularly its posterior aspects" They also note that destruction of PPC in the macaque does not lead to those severe visual defects (especially neglect) experienced by humans by similar loss caused by stroke.

In humans it has been shown that SPL and parts of IPS form the important system for constructing an attention feedback signal (Corbetta et al. 2008; Bressler et al. 2008) as has a similar region in the macaque. Thus we can initially identify in both the human and macaque the inverse model controller (IMC) for creating the attention feedback signal as sited in the IPS/SPL of both primates. One would expect a similar identification for intermediate primates.

From the analyses of (Husain and Nachev 2006) the human SPL is to be identified with the total PPC of macaque, so of its SPL, IPS and IPL. This can be read as saying that the macaque only possesses an inverse model controller (IMC) for its attention control in its parietal lobe; it would thus have no further attention copy apparatus and especially any form of short-term memory for it. Such a reading of the macaque brain could thus have serious implications for the conscious experience of the macaque (and most likely lower animals): *it will have no inner self.* It will possess the ability to 'think' and especially 'reason' to attain its goals, provided it has suitable short-term memory and prefrontal goal handling sites. But all this is expected to be inefficient and most especially have no-one 'experiencing' these activities. The macaque may thus be an example of a living zombie.

The above conclusion could be tested directly to see if indeed, across the single cells of the macaque posterior parietal cortex there was an inhomogeneity of the single cell responses, by extending the analysis. Moreover the timing of these signals is crucial, with early signals from stimulus inputs in parietal (at around 200 mille second post-stimulus) indicating the possibility of an attention copy signal being present. However little is presently known about such timing and signal spatial distribution; high-density EEG measurements might be able to help fill that gap. Already initial EEG data show that there may well be early covert attention control in the macaque from observation of the monkey homologue of the human N2pc (Woodman et al. 2007) but further details about any attention corollary discharge signal are still to be obtained. In any case there is a model of the N2pc which indicates that it arises by a competitive process as if from a saliency map (Fragopanagos et al. 2009), so only corresponds to stages 1 or 2 in the evolutionary analysis of attention in Sect. 12.3.

12.5.6 *The Upper Paleolithic Revolution*

The attention-based approach has further implications for expectations of finding behavioural correlates of bigger brains, especially in humans. One of the current puzzles of paleoanthropology, for instance, is the apparent gap between the appearance of anatomically-modern humans some 100,000 years ago, and the advent of unequivocally modern behavior. The latter is conservatively marked by new lithic blade industries in Europe around 45,000 years ago (and probably somewhat earlier in Africa), and more liberally by the over-water peopling of Australia, at around 60,000 years ago. If function leads structure in brain evolution, then why are modern-looking people with modern-size brains not *acting* modern for some 40,000 years (Klein 1989)?

The period around 40,000–50,000 years ago is termed the Upper Paleolithic Revolution. It involved the rather sudden appearance of important indicators of the beginnings of self-knowledge in humans: self-adornment, the creation of sculptures, cave painting of a quasi-religious nature and the creation of far more sophisticated tools than before. Related to this is the question as to the concomitant disappearance of Neanderthal man around that time.

It has been seriously suggested that there was a fusion of Neanderthal and homo sapiens genes around the Upper Paleolithic Revolution, so causing the vigour of the fused homo after that phase (Cochran and Harpending 2009). However there is still the problem of how the resulting genetic change could have led to the evolution of intelligence as observed in the rapid growth of culture and civilisations during the last 10,000 years, nor does it help explain the Upper Paleolithic Revolution somewhat earlier. However our basic conjecture still stands that it was through the evolution of a suitably powerful attention control system that an inner-self-based consciousness emerged; it could indeed be that the genetic fusion of Neanderthal and homo sapiens helped achieve that evolution of attention.

To return to more factual aspects on the evolution of attention, the cerebellum plays an important role in numerous cognitive as well as motor acts; one observable aspect of that is the reduced ability of those with cerebellar lesions to achieve covert attention control of eye movements: "Mid-cerebellar regions are indirectly involved in covert visual attention via occulomotor control mechanisms" (Bajer et al. 2010, abstract). From the results of (Weaver 2005) on relative cerebellum volumes with respect to that of the overall brain we might expect that there was increase in the ability for covert attention control, with an increase reported by Weaver of the cerebellum of recent man: "An increase in cognitive efficiency as a result of expanded cerebellar capacity in the Late Pleiostocene and Holocene humans" (Weaver 2005, p. 3580). It is that increase, we conjecture, that brought about the Upper Paleolithic Revolution and thence man's present level of consciousness.

One support for this conjecture is that the archaeological evidence for advanced behaviour at earlier time ranges has not yet been found. Of course it may be in due course, but if not then the modification of brain tissue through both a change in the cerebellum/cortical ratios and the expansion of the parietal and prefrontal lobes around 40,000–50,000 BC would have led to greater powers of attention control. Indeed we can conjecture that the differentiation of modern man at the time of the Upper Paleolithic Revolution was due to the increased expansion of attention powers, possible through the ability to use the extra parietal brain tissue for additional attention control through a copy of the attention movement control signal (Taylor 2000, 2007), as aided by increase of the cerebellum.

A further point of support is that the filling out of the full attention circuitry of stage 5 and Fig. 12.5 would require a suitably enlarged brain. As noted earlier the macaque does not have such an enlargement (Husain and Nachev 2006), and only higher primates, especially humans, may reach that stage.

We can only summarise that attention control could have become more sophisticated as the brain evolved, leading eventually to the possibility of a CODAM-style architecture able to support an inner self as having the experience of the sensory input. Thereby humans were rescued from the fate of being zombies, as conjectured above for the macaque. To properly support this conjecture, considerable work needs to be done to develop the evolutionary ladder of attention control in mammals. There is already work indicating the presence of internal (forward/inverse) models for motor control, especially in prey hunting, such as in bats (Erwin 2009; Erwin and Wilson 2001). There is also considerable experimentation on attention in rats, such as the effect of destruction of their nucleus reticularis thalami. However similar results are not available for more general attention control, so that vital information is wanting. At the same time more intensive work needs to be carried out on fossil endocasts of early mammalian skulls, to detect if the parietal (or other) neural modules were modified to take account of parietal (as well as other brain region) extensions expected to handle the expansion of attention control powers related to the emergence of consciousness.

12.6 Infant Attention Development

We can add somewhat to the evidence presented above, especially in terms of the early development of attention in infants and young children. Whilst there is no proof that attention evolved in the same manner there is the usual saying that 'Ontogeny mirrors phylogeny'.

There have been numerous studies of the three stages of reflexive attention control observed by stimuli in the environment, initially at sub-cortical and lower-level cortical brain sites in the infant (Reynolds and Richards 2008). Inner processing up to the level of the temporal and parietal lobes in cortex then develops, and finally there is the use of executive control enabling long-term goals to be attended to. This latter stage is thought to arise due to the beginning to come on stream in the infant/child brain of the prefrontal goal control sites associated with the powers of endogenous attention. The first stage of attention development involves the exogenous control of attention, in a saliency form, as controlling where the infant looks as directed by its environment. There is also, in the earlier period, an increased degree of 'capture' of the infant's centre of gaze by its surroundings, which much reduces as the more endogenous control of attention develops.

Such development supports the earlier gradual evolution of attention control, as proposed in Figs. 12.1, 12.2, 12.3, 12.4, and 12.5.

12.7 Conclusions

The main basis for the evolution of consciousness proposed here is based on obtaining a better understanding of how attention could have evolved. A possible way this could have occurred was proposed in Sect. 12.3 through the gradual evolution of the complexity of the control structure of attention as observed in the adult humans. Experimental support for this evolution-of-attention proposal was given from the following observations:

1. The observation of attention powers in lower level animals, even at the level of invertebrates such as arthropods and cephalopods.
2. The development of attention powers in vertebrates, especially primates, culminating in the attention powers of humans.
3. The development of attention in infants as discussed in the previous section, with the three stages of reflexive attention control by stimuli in the environment by sub-cortical and lower-level cortical brain sites; inner processing up to the level of the temporal and parietal lobes in cortex, and then ultimately the use of executive control enabling long-term goals to be attended to.
4. The sudden surge of 'ownership' knowledge in humans at about the same time as the Paleolithic Revolution, and the co-evolution of the cerebellum as compared to the cortex.
5. The difference observed between macaque and human brains, especially in the parietal lobe, with the macaque missing the expanded lower parietal area of the human brain.

We proposed in this paper the conjecture that 'I' evolved over geological time in order to make the essential attention control apparatus more efficient. It did that by evolving, using the results of items 1–3 above, to provide a copy of the attention movement signal, usually termed a corollary discharge. It was this signal that is suggested as being available in the human as compared to the macaque and other lower brains. This then allows, for humans, attention control so improved as to have the following four evolutionarily-advantageous properties (among others) of:

1. Speed-up of amplification of the target to attain its working memory buffer through attention amplification;
2. Inhibiting distracters from being incorrectly amplified and hence also attaining the working memory buffer;

3. Allowing for error calculation and redirection of the attention feedback signal to remove the error;
4. Providing an owner/centring experience, so giving an initial answer to the hard problem of Chalmers (1996), as well as providing a further answer to the changes behind the Upper Paleolithic Revolution.

Items 1–3 above have experimental justification as noted above; item 4 can be attacked both by experiment as well as through the introduction of aspects of inner experience into our understanding which is inaccessible in animals other than humans.

The discussion of the previous section implies that there is considerable experimental work required, both of comparative neuro-anatomical and paleoneurological form before the route proposed here for consciousness to have evolved in humans is properly supported. However the paper provides a somewhat different approach to the way the task has been studied previously. It focuses on the evolution and comparative anatomy of the attention control system in animals as being the important faculty to probe in order to understand the important evolutionary emergence of consciousness.

References

Allen C (2009) Animal consciousness. Stanford Encyclopedia of Philosophy (online)

Baars BJ (2002) The conscious access hypothesis: origins and recent evidence. Trends Cogn Sci 6(1):47–51

Baars B (2005) Global workspace theory of consciousness: toward a cognitive neuroscience of human experience. Prog Brain Res 150:45–53

Bajer B, Dieterich M, Stoeter P, Birklein F, Muller NG (2010) Anatomical correlates of impaired covert visual attention processed in patients with Cerebellum lesions. J Neurosci 2010(10):3770–3776

Bressler S, Tang W, Sylvester CM, Shulman GL, Corbetta M (2008) Top-down control of human visual cortex by frontal and parietal cortex in anticipatory visual spatial attention. J Neurosci 28(40):10056–10061

Buller J (2009) Evolution of the mind: 4 fallacies of psychology. Scientific American

Butler AB (2008) Evolution of brains, cognition and consciousness. Brain Res Bull 75:442–449

Chalmers D (1996) The conscious mind: in search of a fundamental theory. Oxford University Press, Oxford

Cochran G, Harpending H (2009) The 10,000 year explosion: how civilisation accelerated human evolution. Basic Books, New York

Corbetta M, Patel G, Shulman G (2008) The reorienting system of the human brain: from environment to theory of mind. Neuron 58:306–324

Darwin C (2004, 1871) The descent of man and selection in relation to sex. Penguin Classics, New York

Desimone R, Duncan J (1995) Neural mechanisms of selective visual attention. Annu Rev Neurosci 18:193–222

Erwin HR (2009) M-Systems and consciousness. CASA'09, Liege, Belgium, 3–8 Aug 2009

Edelman DB, Seth AK (2009) Animal consciousness: a synthetic approach. Trends Neurosci 32(9):478–484

Erwin HR, Wilson WW (2001) A computational sensorimotor model of bat echolocation. J Acoust Soc Am 110(2):1176–1182
Finlay BL, Darlington RB, Nicastro N (2001) Developmental structure in brain evolution. Behav Brain Sci 24:298–304
Fragopanagos N, Kockelkoren S, Taylor JG (2005) A neurodynamic model of the attentional blink. Cogn Brain Res 24:568–586
Fragopanagos N, Cristescu T, Goolsby B, Kiss M, Eimer M, Nobre AC, Raymond JE, Shapiro KL, Taylor JG (2009) Modelling distracter devaluation (DD) and its neurophysiological correlates. Neurospychologia 47:2354–2366
Gangestad SW, Simpson JA (2007) The evolution of mind: fundamental questions and controversies. Guilford Press, New York
Geary DC (2004) Evolution and cognitive development. In: Burgess R, MacDonald K (eds) Evolutionary perspectives on human development. Sage Publications, Thousand Oaks, pp 99–133
Hilgetag CC, Lomber SG, Payne BR (2001) Neural mechanisms of spatial attention in the cat. Neurocomputing 38–40:1281–1287
Hochner B, Shomrat T, Fiorito G (2006) The octopus: a models for a comparative analysis of the evolution of learning and memory mechanisms. Biol Bull 210:308–317
Husain M, Nachev P (2006) Space and the parietal cortex. Trends Cogn Neurosci 11(1):30–36
Jaynes J (1976) The origin of consciousness in the breakdown of the bicameral mind. Princeton University Press, Princeton
Jerison HJ (2006) Evolution of the frontal lobes. In: Miller BL, Cummings JL (eds) The human frontal lobes: functions and disorders, 2nd edn. Guilford Press, New York, pp 107–118
Jerison HJ (2007) Evolution of the frontal lobes. In: Miller BL, Cummings JL (eds) The human frontal lobes. Guilford Press, New York
Klein HG (1989) The human career: human biological and cultural origins. Chicago University Press, Chicago
Koch C, Ullman S (1985) Shifts in selective visual attention: towards the underlying neural circuitry. Hum Neurobiol 4:219–227
Levine J (1983) Materialism and Qualia: the explanatory gap. Pac Philos Quart 64:354–361
Lewis-Williams D (2002) The mind in the cave. Thames & Hudson, London
Li Z, Dayan P (2006) Pre-attentive visual selection. Neural Netw 19(9):1437–1439
Nagel T (1974) What is it like to be a bat? Philos Rev 83:434–450
Nieuwenhuys R (1994) The neocortex. An overview of its evolutionary development, structural organization and synaptology. Anat Embryol (Berl) 190(4):307–337
Northcutt RG (2002) Understanding vertebrate brain evolution. Integr Comp Biol 42:743–756
Reynolds GD, Richards JE (2008) Attention and early brain development. In: Tremblay RE, Barr RG, Peters RDV, Boivin M (eds) Encyclopedia on early childhood development [online]. Centre of Excellence for Early Childhood Development, Montreal, pp 1–5
Sumner P, Adamjee T, Mollon J (2002) Signals invisible to the collicular and magnocellular pathways can capture visual attention. Current Biology 12(15):1312–1316
Taylor JG (2000) A control model for attention and consciousness. Soc Neurosci Abstr 26, 2231#839.3
Taylor JG (2002a) Paying attention to consciousness. Trends Cogn Sci 6(5):206–210
Taylor JG (2002b) From matter to mind. J Consciousness Stud 6:3–22
Taylor JG (2007) CODAM: a model of attention leading to the creation of consciousness. Scholarpedia 2(11):1598
Taylor JG (2010a) The creativity effect: consciousness versus attention. In: The 2010 International Joint Conference on Neural Networks (IJCNN), Barcelona, IEEE, pp 1–8, 18–23 July 2010
Taylor JG (2010b) Consciousness versus attention. In: International conference on artificial neural networks; artificial neural networks – ICANN 2010. Lecture Notes in Computer Science, vol 6354, pp 496–503
Taylor JG (2010c) The I's eye view of its consciousness. J Consciousness Stud 17(1/2):95–117

Taylor JG (2010d) A neural model of the loss of self in schizophrenia. Schizoph Bull (on-line: 23 April 2010)
Weaver AH (2005) Reciprocal evolution of the cerebellum and neocortex in fossil humans. Proc Natl Acad Sci 102(10):3576–3580
Woodman GF, Kang M-S, Rossi AF, Schall JD (2007) Nonhuman primate event-related potentials indexing covert shifts of attention. Proc Natl Acad Sci U S A 104(38):15111–15116
Xu Y, Chun M-M (2006) Dissociable neural mechanisms supporting visual short-term memory. Nature 440:91–95

Chapter 13
Animal Consciousness

13.1 We Are All Little Lambs?

We know far too little about whether or not animals are conscious. Of course it is usually said that there are a range of consciousness that we should consider and that at a lower level than ourselves, it is granted, an animal might possess consciousness. It seems that many an animal-lover believes this possibility. So far we have not considered different levels of consciousness, but only human consciousness. It is important to ask if there could be other levels of consciousness besides that of human consciousness. To prepare for that we also need to consider what levels of intelligence these animals can reach. I will start with a personal story indicating that it is not only animal consciousness but also animal intelligence that we do not understand very much.

My daughter lives in Northumberland with a host of dogs, birds and other animals as well as two children and a veterinary husband. My wife and I arrived there just before last Christmas (2011), where we found the kitchen (the warmest room in the house, with a 24-h Aga Cooker) to be somewhat taken over by two adult Labradors, one puppy Labrador, various gerbils, and a woolly black lamb (a Suffolk, called Charlie). Such a concatenation of animals was initially quite overpowering, but the novel addition was the black lamb. Charlie was rather beautiful, and had been included in the kitchen's select clientele due to his precarious health. As the days passed Charlie became more and more intertwined with the dogs, so that within a few days he was part of the huddle of bodies soaking up the warmth in front of the Aga during the sleeping periods of the animals.

I watched him over the next 10 days develop both physically as well as in relation to the dogs and their toys. For Charlie became part of the dog pack not only huddling down in a pile with the others, but also learning to play with the dogs' special toys, especially those of the puppy. He naturally had difficulties since his little hoofs were not as flexible as a puppy's paws, and although Charlie watched with care he was not able to perform most of the manipulations of the puppy. Yet he did persevere.

J.G. Taylor, *Solving the Mind-Body Problem by the CODAM Neural Model of Consciousness?*, Springer Series in Cognitive and Neural Systems 9,
DOI 10.1007/978-94-007-7645-6_13,

He also looked around the kitchen and found his own niche of toys (my daughter's cookery books, which he was able to knock off a low bookshelf) and that kept him very happy and occupied.

The above example shows that an animal like a lamb can be brought up in an enriched environment so as to also enrich its own life. Extending this to many animals and their progeny is clearly impossible due to the cost. Even Charlie will have his comeuppance when he is put out in the fields with the other sheep and lambs. Are they all living reduced lives, with a limitation artificially imposed by their environment on the possible levels of cognition they could achieve in their lives? And is this true across a large range of animals?

Some animals show their intelligence in a variety of ways, and give a strong demonstration of their power. It was exactly this that the dolphin Kelly, living in Mississippi, did recently. She worked out it was better to hide pieces of rubbish (for which she was rewarded if she gave them to her trainer) and tear off pieces whenever she was hungry; in this way the rubbish went much further! Discovery of this sophisticated strategy has even led to a call at the 2012 American Association for the Advancement of Science to consider whales, dolphins and porpoises as 'non-human persons'. Perhaps. But however, the claim that "Science has shown that individuality, and self-awareness is no longer a unique human property", as stated by the philosopher Thomas White of Loyola Marymount University in Los Angeles, goes too far. It is true that beyond humans, elephants and great apes, as well as certain cetaceans, are able to recognise themselves in a mirror. But, according to CODAM, that is not enough – they must also possess a complex enough brain architecture to possess a pre-reflective self. We shall see what science can say about such a claim shortly.

13.2 Animal Cognition

One of the features of the understanding of animals over the recent decades is that of their cognitive powers. Careful research into a number of animals, by observation of those living either in the wild or under laboratory conditions, has led to a drastic revival as to what these cognitive powers really consist of.

Thus elephants are noted for their memory ('an elephant never forgets' is a well-known saying). This memory faculty is very important for them to survive, since they can remember sources of water and food, as well as transmit such knowledge to others of their tribe. Another example is that of passing on of knowledge by meerkats to each other of how to manipulate dangerous prey. Moreover adult meerkats will bring intact live scorpions with their stings bitten off to their offspring. Bees, ants and some carnivores are also known to actively pass on knowledge of this sort.

On the aspect of emotions in animals, crows have been recognized as feeling fear. Interestingly in relation to the previous chapter on emotions, a scan of the crow's brain in interaction with scientists who have ringed them in the past, showed that the crow's amygdala was active when they saw the face of their captor (which they recognized from the past).

Of course we expect chimpanzees to have some high level cognitive powers. A nice case of this was reported recently in The Times (Bonnin 2012): "I traveled to Kyoto University in Japan to compete against Ayumu, an 11-year old chimpanzee, in a memory test. Nine numbers would flash up for 60 ms, faster than the blink of an eye, all that we had to do was recall them in sequential order. I failed miserably. Ayumu passed with flying colours." So humans beware!

In particular, reasoning – which we define as the drawing of conclusions from facts – is one of the most important faculties possessed by humans, but now is also thought by some researchers to be used, at a lower level, by chimpanzees and other primates and even by New Caledonian Crows, although that is also still controversial. In this chapter we will develop a simple neural model of reasoning at a non-linguistic level for tasks that have been heavily used in animal cognition research, such as on those animals mentioned above. The non-linguistic aspect of the reasoning considered here denies the use of language, so of any logical reasoning. At the same time it reduces the range of tasks that can be reasoned about to those at an instrumental level.

As examples of tasks set to animals, the trap-tube, food extraction and 2-stick tasks represent tasks in which the animal has to try to obtain food by suitable manipulation of objects to which they are already accustomed. Thus the trap-tube task requires the animal to extract a piece of food from a horizontal tube which has a vertical trap somewhere along its length. A stick is provided to push the food out of the tube, but to be successful that action can be done only from one end, that on the opposite side of the tube to the hole. The food extraction task requires an animal to remove a basket of food from a transparent vertical tube using a long wire with a hook at one end. Finally in the 2-sticks task an animal needs to use a short stick, which is at hand, to draw to itself a longer stick (out of direct reach) so that it can then acquire a piece of otherwise inaccessible food.

The reasoning elements in the above tasks are relatively clear, with each of them able to be solved by a suitable strategy. However there is considerable controversy as to these tasks being able to be solved by simpler non-reasoning methods, such as by chaining (sequence learning).

There is also recent controversy as to the extent that birds, for example, possess any cognitive powers and at what level these powers are expressed in any given species. Thus in a searching review on the cognitive powers of New Caledonian crows, Professor Kacelnik of Oxford University and his colleagues concluded recently that "In our opinion, claims for analogical reasoning based upon sequential tool use remain unjustified, and using sequential tool use as a benchmark of this ability is inappropriate. Reasoning (let alone analogical reasoning) is not the only cognitive mechanism to account for sequential tool use: simpler processes such as chaining may be sufficient, as noted above." (Wimpenny et al. 2009, pp. 11–12). In particular they add in the abstract that "By analyzing tool choice, tool swapping and improvement over time, we show that successful subjects did not use a random probing strategy. However, we find no firm evidence to support previous claims that sequential tool use demonstrates analogical reasoning or human-like planning." Thus there seems to be no evidence of trial and error learning neither does there seem to be any analogical reasoning but there is reasonably powerful use of some form of processing based on sequence learning.

On the other hand, a somewhat different conclusion is reached in another recent paper by a different group (Taylor et al. 2009), who conclude in their abstract "Causal and analogical reasoning may form the basis of the New Caledonian crow's exceptional tool skills." However we have to be careful in making such a comparison since the latter group were analysing the effectiveness of New Caledonian Crows on the trap-tube paradigm whereas the former (Wimpenny et al. 2009) were concerned with the effectiveness of such crows in tasks based on multiple tool use.

At the same time there is also controversy over reasoning in other animals, especially in primates such as chimpanzees and their close relatives (bonobos, gorillas and capuchin monkeys, for example). Thus it seems that some form of analogical reasoning is observed in chimpanzees, as reported earlier by (Gillan et al. 1981) for a chimpanzee able to reason through use of learnt symbols. These authors studied analogical reasoning in a 16-year-old female chimpanzee (Sarah) in several experiments using the $A - B$ same as $A' - B'$, where 'same as' was a specific symbol taught to Sarah. The general design of the analogy problems was "A B'," where 'same' was Sarah's plastic symbol for it. Sarah solved analogy problems with two types of displays: (a) forced choice, in which she had to complete an analogy by choosing the correct B' from a set of alternatives and (b) same–different, in which she had to complete an analogy by choosing the correct predicate, same or different. In addition, she correctly solved both figural analogy problems, in which the stimuli were geometric figures and the relations among them were differences in size, color, or marking, and conceptual analogy problems, in which the stimuli were household objects and the relations were functional and spatial. The data indicated that Sarah used the relation between A and A' to solve the analogy problems.

Since that time there have been many analyses of such reasoning powers in primates, both intrinsically and comparatively. The situation is still controversial, however, especially for their powers of understanding the motives and issues of other con-specifics, and as (Povinelli and Vonk 2006) point out: "The question of whether chimpanzees, like humans, reason about unobservable mental states remains highly controversial. On one account, chimpanzees are seen as possessing a psychological system for social cognition that represents and reasons about behaviors alone. A competing account allows that the chimpanzee's social cognition system *additionally* construes the behaviors it represents in terms of mental states. Because the range of behaviors that each of the two systems can generate is not currently known, and because the latter system depends upon the former, determining the presence of this latter system in chimpanzees is a far more difficult task than has been assumed. We call for recognition of this problem, and a shift from experimental paradigms that cannot resolve this question, to ones that might allow researchers to intelligently determine when it is necessary to postulate the presence of a system which reasons about both behavior *and* mental states."

13.3 The Use of Internal Models

Alternative support for animal reasoning, enabling a more precise view of its mechanism and hence its detailed nature, will arise from extracting from the brain the mechanisms that are behind any supposed cognitive powers that an animal might possess. This can be regarded as adding further difficulty to the search for animal cognition, especially in birds, for which the nature of brain processes is still more poorly understood than in primates. Yet such a study of brain mechanisms must ultimately be achieved, since it is clear that it is in the brain of an animal that such cognitive powers reside. Knowing how such mechanisms are achieved will enable us to decide what sort of cognition is occurring, be it analogical reasoning, simple association, and about others' mental states or whatever. In this chapter we will consider mainly the possible reasoning powers of an animal and the underlying brain mechanisms that could be involved, when apparent analogical reasoning seems to be occurring.

In order for an animal to reason, we propose that the basis of any reasoning and other cognitive powers possessed by the animal brain is by means of suitable internal models of the world. These internal models consist of three crucial components: (1) Forward models FM (coding causality of the world); (2) Working memory (buffer) modules WM (so as to aid imagining/estimating the world by holding activity over suitable lengths of time for them to be involved in further processing); (3) Inverse model controllers IMC (which generate desired control actions). On top of all these a filter control system is needed which divides the world into a 'to be attended to' component and a 'remainder', equal to that part composed of distracters. The FM/IMC pairs are a standard part of modern control theory. However such an approach has no WM or attention components, especially in neural form.

The presence of a forward model seems essential for any animal to be able to work out the factual consequences of an action when it is in a given state, without performing that action. Using a forward model would lead to the animal being able to deduce a conclusion (the new state it will be in) from facts (these latter being the state the animal is in initially and the proposed actions to be taken). That may be regarded as the simplest form of reasoning, following our earlier definition. Another form of reasoning would be through an inverse model, producing a conclusion – the action needing to be taken – in order to move from a given state to a desired one. Further chaining or sequencing would be necessary for more complex situations, so agreeing with the quote from (Wimpenny et al. 2009). But the most basic feature of reasoning, if defined as we just have above, is provided for by the presence of a (set of) forward and inverse models in the animal's brain and working memory sites to enable the resulting activities of the inverse and forward models to be used in sequences or other situations or just be held till the results of such a reasoning process is completed so that actual actions can be taken.

There has been much discussion in the literature of internal models used in motor control (Kawato and Gomi 1992; Wolpert et al. 1995; Kawato 1999; Miall and Wolpert 1996; Desmurget and Grafton 2000; Wolpert and Gharharmani 2000). This has been extended to coupled pairs of FM/IMC models in the MOSAIC architecture and its extensions (Wolpert and Kawato 1998). Such internal motor models have been applied to the learning of intentions of others in (Oztop et al. 2004). These models are proposed to generate fast prediction processes, by means of efferent copies of motor actions, used to speed up responses and reduce or eliminate errors in performance. Here we use this class of models to generate by recurrence virtual prediction processes to enable the ascertaining of how to attain desired goals. We do so using additional values attached to the stimuli observable in an animal's external environment, as well as the working memory structures mentioned above.

It is appropriate to note more specifically here that the internal models considered by others and noted in the previous paragraph are used, in the main, in a different arena and for different functions than to be at the basis of reasoning, as considered here. The MOSAIC class of coupled forward/inverse models (Wolpert and Kawato 1998) is similar to those postulated here for reasoning purposes; they are an ensemble of inverse/forward models for motor control and planning. However they have a specific mechanism for choice (termed 'responsibility') of a given pair of models to use in a given situation in (Wolpert and Kawato 1998) and later applications. This is based on using the forward models to generate predictions leading importantly to error levels and so to the responsibility component. We are interested here in animal reasoning, so in using a suitable ensemble of internal models to make covert predictions of the future effects of motor responses (as generated by the coupled internal/forward models for a given context). This is carrying the models into the future inside the animal's brain without actions being made, so requiring some form of working memory to hold the results of partial forward planning.

There is a more important difference to MOSAIC that we introduce: for an animal the most important aspect of its life is continued survival, attained by searching for and gaining appropriately rewarding goals, such as food. It is this use of rewarded goals, and even their reward manipulation which will lead us to a different architecture than that of MOSAIC (with its responsibility determined by a somewhat non-biological soft-max function).

How will this type of architecture help to support animal reasoning? Let us be clear what it is we expect from animal reasoning, as well as develop some specific examples. A general definition we have used already is that 'reasoning is the drawing of conclusions from facts'. This would seem to suit most animal reasoning processes that might occur since the conclusions that animals reach in terms of their putative reasoning is about the action to be taken to achieve a certain goal, usually (but not always) of a food nature. That is where the need for an inverse control model – generating the desired action – is clear in the reasoning animal's brain. Powers of causal understanding (Shettleworth and Sutton 2006), gaze following and competing for food (Tomasello et al. 2007) are also recognised in animals. In all of these analyses and paradigms the animal brain appears to have abilities which would be ascribed to reason if they occurred in observed human behaviour.

We are then faced with the following question: How does the animal brain reason (if it does) if it has no power of logic, so that it cannot use the deductive power of propositional logic? It would appear only to be able to use inductive or abductive reasoning methods, at least according to the standard definitions of reasoning (where inductive reasoning conclusions are only drawn with a certain probability, and abductive reasoning involves inference to the best explanation). However we are crucially concerned with determining the correct actions to achieve certain well-defined goals, so with what is currently termed 'Practical reasoning'.

Much discussion of practical reasoning has occurred as part of ethics (Hume 1777/1975; Millgram 1997). Here we wish to consider practical reasoning as part of the more general area of cognition. This is especially so since we are considering animal powers in preference to human ones, so ethical aspects may not be so relevant.

As an example of the practical reasoning aspect, let us consider the trap-tube task. Here the animal has to work out the actions required to obtain the food in the tube by using a stick (a tool it has used from an early age) but at the same time avoid losing the food in the vertical trapping tube. The action required to achieve this is not known ab initio by the animal: it has to be worked out by it in one way or another. If the animal makes a lot of random-appearing moves until it succeeds then it is using trial-and-error learning to achieve its end goal. However if the animal remains motionless for a while (although this may not be for long, so could be difficult to measure by pure behavioural assessment) and then makes a series of correct moves to achieve its goal it would be expected that some form of reasoning will have been used by the animal. If the animal makes the set of actions to achieve its goal immediately it is in the task situation, then we expect that it has already met this situation before and is using its memory to be successful. In either case we can ask as to what sort of mental processes would have been occurring in the animal's brain in this latter case.

13.4 Observing Reasoning by Internal Models

Reasoning, along the lines we have defined it of 'drawing a conclusion from facts', could be obtained by possession, in the animal's brain, of suitable models of the world that allow associative recall to be used in a contextually sensitive manner. We can consider such processing as 'guided chaining', where the guidance arises from the context, be it from the environment or by means of an action being taken as determined by the state it is in and the state it wants to get into. The neural states being used in this processing need not be linguistically encoded, so explaining how non-linguistic reasoning could arise.

If we consider 'guided chaining' in more detail, then an animal will require some form of internal model to achieve it, by use of either an FM or an IMC. For it cannot go from one visual state to another, for example when the animal is thinking of them, without use of a suitable action to make the transfer; nor can it make a required action to a desired state without a suitable inverse model. Since these are effectively all the changes an animal can make, then we need the IMC/FM pair to achieve any primitive reasoning; this is what we have just claimed.

One feature that we must consider is that of being off-line or on-line in using the internal forward model (by off-line we mean using the internal model in a silent or non-active mode). Reasoning itself, as an activity, would be expected to involve the off-line mode. Otherwise an animal could launch itself into an unrewarding, if not dangerous, situation. However the on-line (or immediate response) mode would be possible if the animal had suitably memorised the result (say by chaining). It could then make a more rapid response, if any action was needed. The off-line use of internal models (so as a reasoning process) is something that would be expected to require little immediate response, so the animal would be somewhat more stationary than normal. Of course the animal could just be standing still for a rest or for a different reason than that of reasoning; it could also be that the amount of time needed to perform any reasoning could well be rather too brief to be measurable for the animal. Thus it is possible that such reasoning might only be detectable by brain measurements during the overall response period to check that, for primates, the relevant brain regions (at least those containing internal models) are active. For birds it would be necessary to understand at a much deeper level than at present what is happening in the bird brain during many aspects of information processing, but most especially in the run-up to its making successful actions in the tasks mentioned above (and others).

We have already suggested that an animal reasons by the off-line use of triplets of internal models FM/IMC/WM (with additional attention components to enable the overall system to resist distracters). This fits into what we have just described about reasoning, since a forward model produces a new state from a previous one by a suitably chosen action, with a context determined by the action taken.

13.5 Animal Consciousness?

Do some animals have the further power of being conscious? From our discussion of the evolution of consciousness (in Chap. 12) we recognised this as depending heavily on the powers of attention to which a given animal had evolved. We even had some discussion in that chapter of possible animal consciousnesses. We especially considered primates with their very efficient attention apparatus, although possibly still limited through evolution in not yet reaching the possibility of implementing attention at the CODAM level. A specific example of this was given as the macaque, where it was noted that there is tentative evidence for the macaque parietal lobe being crucially smaller than that of the human.

But what might we expect in some form of reduced consciousness in animals? So far we have considered only either CODAM or the reduced class of attention control systems in Chap. 12, as well as our original discussion of attention control in Chaps. 5 and 6. Is it not possible that there is a continuum of attention control systems, each melding into the next higher up in the evolutionary process as the modules of the previous level have expanded? So there would at least be a continuum of attentions, although it is unclear that would lead to a continuum of consciousness without some further analysis.

Here we must make it clear what it is we are taking consciousness to consist of, in either animal or human. Through CODAM and the many discussions associated with the ownership characteristics, we will only accept a conscious system as one which has an owner. This is a very important characteristic. Without it there can be no 'I', as we have discussed many times so far in this book. It is this aspect, of there 'being someone there' that is absolutely basic to our notion of consciousness. Without it there may be considerable cognitive level processing, but no-one for whom such processing is occurring.

It is clear we have a serious difficulty in testing animals, if we have had such difficulty as described already in Chap. 9 of proving the existence of a corollary discharge of attention in the human case. But we can see that, given our definition of attention, the required corollary discharge signal is either there (possibly with a low level of distracter inhibition by the corollary discharge) or it is absent. We could expect that for all animals with an attention movement generator (IMC) there would be some leakage from it to be able to act, even at a low level, as a corollary discharge. However there are conditions on this leakage which make it quite non-trivial, since it needs to be able to inhibit distracters. We would expect this to require a suitably sophisticated attention apparatus. A simple leakage from the IMC is clearly not enough to achieve that.

I can only conclude from this discussion that consciousness may indeed be very rare. The claimed consciousness of one's pets (cats or dogs, say) is thus to be regarded as a very sophisticated form of extended ballistic attention control. Charlie, the Suffolk lamb mentioned at the beginning of the chapter, is certainly out of the running for consciousness. Even the macaque has no consciousness, according to the discussion in Chap. 9. That does not say anything about higher primates, the chimpanzee, bonobo, gorilla and orang-utan. They still need to be more carefully investigated, especially through their attention apparatus. This also applies to the claims mentioned at the beginning of the chapter about whales, dolphins and porpoises. There is still a lot of investigation to be carried out on animal attention.

One important warning to be attached to the results here is that they depend most crucially on one experimental result about the macaque parietal lobe compared to that of the human. So we should wait till there is considerably more data available on the brain structures of the relevant animals, especially of their attention control system.

Such is the case for animals. However birds have a rather different brain structure from animals, with only a one layer cortex. At the same time there are other differences between animal and bird brain components. Therefore further study needs to be made in the case of birds.

13.5.1 Animal Welfare

Our main conclusion is that animals below primates may have extensive attention and reasoning apparatus, but do not possess an inner self. They are not conscious in my use of the term. This does not mean that such lower animals should not be dealt with at a suitably humane level of animal welfare. They can detect and enjoy as good a quality of life as is possible. This is in support of the presence of regulations in farming which prevent obnoxious battery farming for hens or sow pens (as recently brought in by the European Union). Nor should it allow distasteful and barbaric practices in killing wild animals. They are all sensitive although, we have concluded, in a different manner from ourselves. They may well be zombies, but with 'feeling' hearts (although without consciousness)!

References

Bonnin L (2012) Beastly brain boxes. The Times, Wednesday, 8 Feb 2012, Times 2, p 17

Desmurget M, Grafton S (2000) Forward modelling allows feedback control for fast reaching movements. Trends Cogn Sci 4(11):423–431

Gillan DJ, Premack D, Woodruff G (1981) Reasoning in the chimpanzee. 1. Analogical reasoning. J Exp Psychol Anim Behav Process 7(1):1–17

Hume D (1777/1975) An enquiry concerning human understanding. In: Selby-Bigge LA, Nidditch PH (eds) Enquiries concerning human understanding and concerning the principles of morals. Oxford University Press, Oxford

Kawato M (1999) Internal models for motor control and trajectory planning. Curr Opin Neurobiol 9:718–727

Kawato M, Gomi H (1992) A computational model of four regions of the cerebellum based on feedback-error learning. Biol Cyber 68(2):95–103

Miall RC, Wolpert DM (1996) Forward models for physiological motor control. Neural Netw 9(8):1265–1279

Millgram E (1997) Practical induction. Harvard University Press, Cambridge, MA

Oztop E, Bradley NS, Arbib MA (2004) Infant grasp learning: a computational model. Exp Brain Res 158(4):480–503

Povinelli DJ, Vonk J (2006) We don't need a microscope to explore the chimpanzee's mind. In: Hurley S (ed) Rational animals. Oxford University Press, Oxford, pp 385–412

Shettleworth SJ, Sutton JE (2006) Do animals know what they know? In: Nudds M, Hurley S (eds) Rational animals. Oxford University Press, Oxford, pp 235–246

Taylor AH, Hunt GR, Medina FS, Gray RD (2009) Do new caledonian crows solve physical problems through causal reasoning? Proc Biol Sci 276(1655):247–254

Tomasello M et al (2007) Reliance on head versus eyes in the gaze following of great apes and human infants: the cooperative eye hypothesis. J Hum Evol 52:314–320

Wimpenny JH, Weir AA, Clayton L, Rutz C, Kacelnik A (2009) Cognitive processes associated with sequential tool use in New Caledonian crows. PLoS One 4(8):e6471. doi:10.1371/journal.pone.0006471

Wolpert DM, Ghahramani Z (2000) Computational principles of movement neuroscience. Nat Neurosci 3:1212–1217

Wolpert DM, Kawato M (1998) Multiple paired forward and inverse models for motor control. Neural Netw 11(7–8):1317–1329

Wolpert DM, Ghahramani Z, Jordan MI (1995) An internal model for sensorimotor integration. Science 269(5232):1880–1882

Chapter 14
Understanding Consciousness and Emotions

14.1 Introduction

There have been many attempts to understand the emotions. In particular there has been a strong thrust towards what has been termed 'affective computing', where human emotion is monitored by a computer, or emotional responses to human activity are attempted to be incorporated in a computer system, an avatar or a robot. But in order to achieve 'affective computing' it is necessary to know what is being computed. That is, in order to compute with what would pass for human emotions, it is necessary to have a computational basis for the emotions themselves. What does it mean quantitatively if a human is sad or angry? How is this affective state computed in their brain? How are emotions 'felt' in the consciousness system? It is these questions, on the very core of the computational nature of the human emotions, which is addressed in this chapter. A proposal will be made as to this computational basis based on the well established approach to emotions as arising from an appraisal of a given situation or event by a specific human being. Finally how emotions can become conscious will be discussed at the end of the chapter.

Previous research in psychology and neuroscience has strived to reach the basis of the emotional process, but the resulting models and theories have often not been translated into a computational representation. In the literature, we can distinguish between those theories advocating a basic emotions approach, a dimensional approach and an appraisal based approach.

In the basic emotion theories (e.g. Plutchik 2001; Panksepp 1982), it is assumed that different processes underlie a small set of basic emotions; more complex emotions would arise from subtle variations on these basic ones. If this were represented computationally, the different basic emotions could have differing and separate underlying computational systems. In this vein, a computational model of fear (see Fig. 14.1) has been proposed by Armony et al. (1995), based around the amygdala as a central structure in fear production, assuming that fear is caused specifically by a (potential) threat and different structures would be responsible for producing other emotions. This model is based on fear conditioning research by

J.G. Taylor, *Solving the Mind-Body Problem by the CODAM Neural Model of Consciousness?*, Springer Series in Cognitive and Neural Systems 9,
DOI 10.1007/978-94-007-7645-6_14, © Springer Science+Business Media Dordrecht 2013

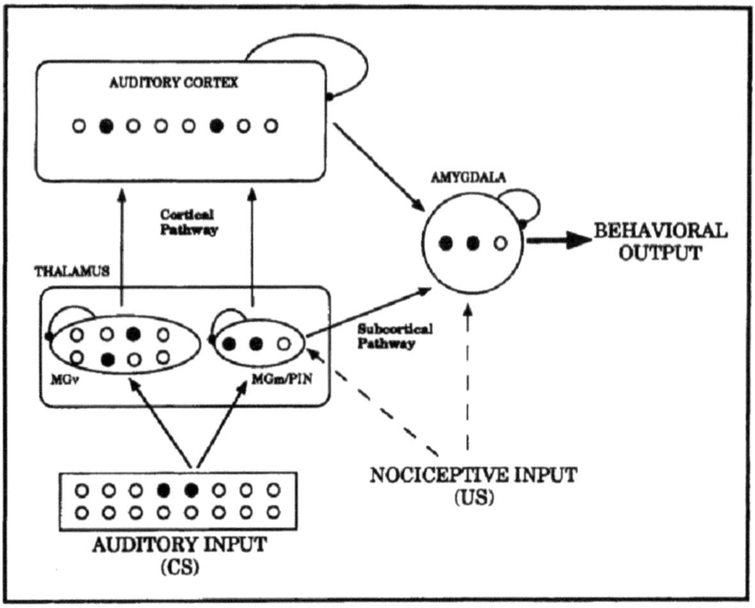

Fig. 14.1 Outline of the fear conditioning model by Armony et al. (1995) (Adapted with permission)

LeDoux (1992), in which it was shown that the amygdala is particularly active in this process. Although this model reproduces experimental results in detail, such specific neural modules have never been located for other emotions. It has later been proposed that the amygdala functions as a relevance detector (Sander et al. 2003) – where stimuli evoking fear tend to be highly relevant.

Dimensional theories state that an emotion is defined by its location in a multidimensional space, where the number and character of the dimensions varies, but tends to include some variation on valence (positive – negative) and arousal (high – low). Computationally, this would translate to a separation between different modules for the different dimensions, rather than a separation between the different emotions, where the output would not consist of categorised emotions but rather of gradual changes in emotional feelings and behaviour. Although it seems that neural responses to valence and arousal can indeed be dissociated (Grimm et al. 2006; Anders et al. 2004), results are not so clear cut that it is possible to point to delineated neural substrates. Nevertheless, psychological research using a statistical technique to identify the principal contributors to a particular phenomenon (such as Principal Components Analysis) has pointed to the existence of precisely four dimensions (Fontaine et al. 2007).

A combination of these two approaches can be found in the work of Russell and Barrett (1999), where core affect is proposed as a two dimensional, permanently present emotional state that does not need to be directed at any particular object or event, whereas prototypical emotional episodes are event-driven and form categories. No explicit computational account is given of this theory, but it suggests some sort of sum or multiplication of the outputs of the dimensional and basic emotion systems as described above results in the overall complex set of emotions.

A more causal approach to the emergence of emotions, which is termed appraisal theory, states that emotions are a function of the individual's interpretation of the situation in terms of (potential) harm or benefit to this individual. A recent detailed account of appraisal theory is proposed in the Component Process Model (CPM) (Scherer et al. 2001; Sander et al. 2005), in which four different Stimulus Evaluation Checks (SEC) – each consisting of several more specific checks – produce different changes in the emotional state. In this account, the SEC's (composed of Relevance, Implication, Coping potential and Normative Significance) are sequential, and each SEC can produce output to various neural systems. As such, a corresponding computational model would contain four modules that are connected in sequence, which each receive input from and produce output to several other neural systems. In such a system, there would not be any one specific emotional output; rather, the emotion would be dispersed across a range of neural systems, most of which would not be specifically 'emotional' in nature.

Ortony et al. (1988) have presented a more inherently computational model of the appraisal process, in which emotions are considered to be valenced reactions to either consequences of events, actions of agents or aspects of objects, whereby other different factors (such as desirability and agency) further discriminate emotions within these categories. A computational structure is clear from the outline of the model in Fig. 14.2. However, there is no connection to potential neural correlates in this model. It should be noted that there is an assumption in this approach that the emotional outputs are categorised, whereas in the CPM this is explicitly not the case.

Thus, quantitative analyses of the emotional processes are still scarce, despite the fact that many theories have been presented that would lend themselves to a quantitative interpretation. In the model we will present in the following sections, we have sought to fill this void in a manner interpretable from a neuroscience as well as a psychological background. The model we present incorporates aspects of the various computational structures outlined above and, as such, forms a blend of their associated theories.

14.2 Basis of Value Comparisons

Our proposal as to the computational origin of emotions is based on appraisal theory, as the emotional output relies on a set of continuous evaluations of the situation. We propose these evaluations to consist of value assessments – both current and predictive of future value. The difference is that where appraisal theorists assume this to be, at its core, a complex cyclical process taking place across different neural systems, a simple core computational basis of the emotional response is proposed

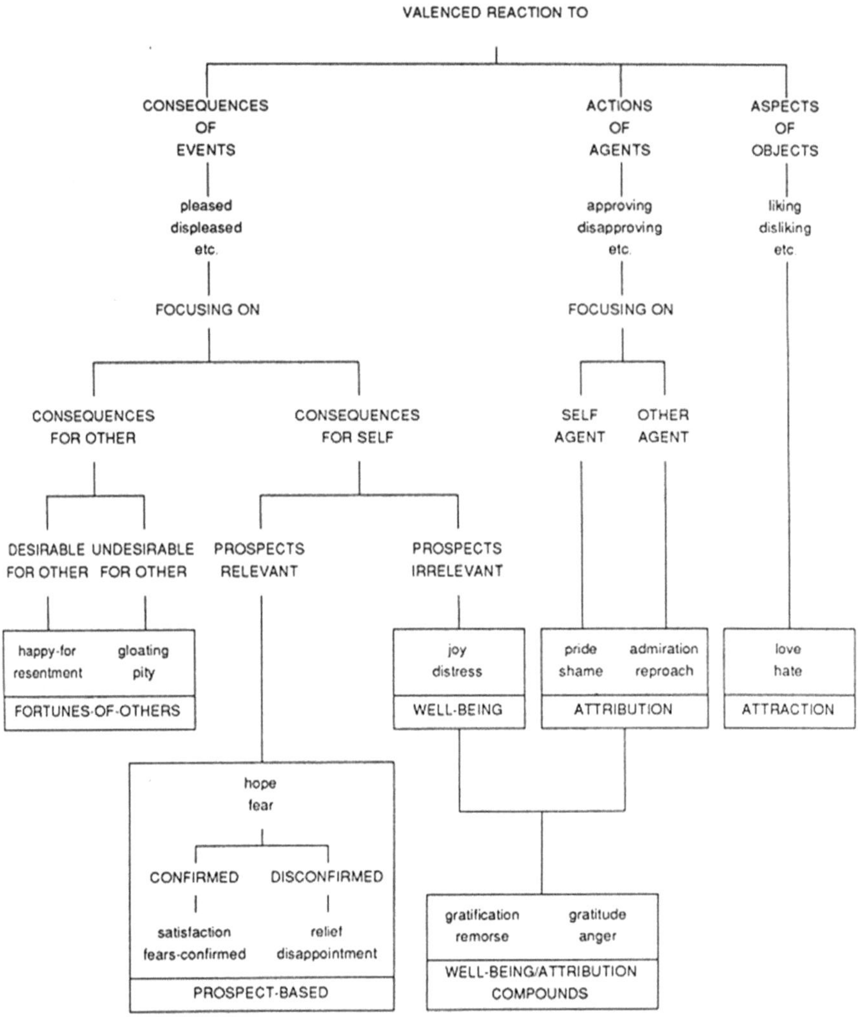

Fig. 14.2 Model by Ortony et al. (1988) (Adapted with permission)

here. This proposal, first presented in Korsten (2009) and also developed in (Korsten and Taylor 2010), is based on value as a core concept that has been related to emotion in psychology as well as neuroscience research.

In neuroscience, research related to value learning has been focused on the process of conditioning, in which an unconditioned stimulus (US) – a stimulus with an intrinsic positive or negative value to the subject – and a conditioned stimulus (CS) – a stimulus that is not intrinsically valuable to the subject – are associated through repeated concurrent or successive presentations, after which responses previously shown to the US only (such as freezing or approach behaviour) are also shown when the CS is presented. This is referred to as a conditioned response (CR). Thus, the CS adopts the value of the US by association. Although this is not an exclusively emotional process (it can occur automatically and subconsciously, and the UR can be non-emotional) the UR tends to consist of behaviours associated with emotion. Therefore, this provides a strong connection between emotion and value in the neuroscientific literature, as the neural basis of this process has been investigated thoroughly and outlined computationally (e.g. Suri and Schultz 1999; Armony et al. 1995). The dopamine signal that has been found to function as a teaching signal in the storage of these associations in particular has been subject to much investigation, but some evidence has also been found for the involvement of the orbitofrontal cortex as the location of long term storage of these associations (Schoenbaum and Roesch 2005), and involvement of the amygdala in the process (see also Fig. 14.1 on fear conditioning).

Appraisals – that is evaluations of the situation, essential to appraisal theory – could be seen as values indicating ongoing or predicted beneficial or harmful effects of the stimulus in the process of being experienced. If we take fear, for example, this emotion could arise from the perception of a stimulus predictive of events that could be physically harmful to the individual. As such, the predicted value of events to come is, from this stimulus, negative; the individual has negative expectations in response to this stimulus. Thus, if we assign a numerical value to this expectation (the expected value) and compare it to whatever we consider a standard, normal value for our physical wellbeing (the normal value) we will find it to be lower. This discrepancy between the expected and normal values can be seen as computationally analogous to the emotion of fear. Similarly, other discrepancies can represent other emotions. Joy, for instance, represents a situation where the current state of events (actual value) is higher than normal, and disappointment represents the assessment that previous expectations (the second expected value) are not fulfilled. In these examples, four different value assessments have been involved:

- Actual value (act), representing the current state of affairs
- Expected value (exp1), representing a prediction based on the current stimulus
- Second expected value (exp2), representing a prediction based on previous stimuli
- Normal value (norm), representing a standard or norm for this value

In the following sections, we will show that discrepancies between these four values can differentiate between a large range of emotions and as such provide a quantitative framework for emotion production.

14.3 Individual Comparisons

One subset of emotions can arise from the physical state – and predictions as to the physical state – of the individual, and can be clearly identified in animals as well as humans. Beyond them are those more specifically recognisable as aroused in

Table 14.1 Value comparisons for the reward/punishment value type

	Reward/punishment
exp1 < norm	Fear
exp1 > norm	Hope
act < norm & exp1 > act	Anger
act < norm & exp1 =< act	Sadness
exp2 < norm & exp1 > exp2	Relief
exp2 > norm & exp1 < exp2	Disappointment
act > norm & exp2 < act	Surprise
act > norm	Joy

social situations, such as shyness or gratitude, or involved with the outcome of some action, as in guilt or regret. We will start by considering the emotions as arising from certain combinations of the four values mentioned at the end of the previous section reaching specific thresholds. We will then expand these processes to apply to the broader range of emotions by considering other aspects of valuations. We present in Table 14.1 putative threshold constraints for the arousal of the corresponding emotion.

The table illustrates how, through the assessment of different values, this framework could differentiate between an even larger range of emotions. We can justify the assignments in the table in more detail as follows:

Fear. This emotion can be thought of as a prediction of a future event that is negative when compared to normal circumstances. For example, a subject seeing a big, strong man holding a stick looking angrily in his direction may predict being beaten up and hurt, and therefore fearful. This is a prediction of the subject's physical comfort being lower than it would be in average normal circumstances. Barr-Zisowitz (2000) notes that: "For the most part, there is a consensus [in the literature] that sadness is distinguished from fear by being a response to an event that has already taken place, whereas fear anticipates an event to come." In terms of value comparisons, this equates for fear to the first expected value (expectation for the future) being lower than the normal value: exp1 < norm.

Hope can be seen as the opposite of fear, so as a prediction that a future event is positive when compared to normal circumstances. This would occur when a subject is presented with a stimulus which would lead them to expect a reward in the near future, such as cooking smells and noises emanating from a kitchen. This equates to the first expected value (expectation for a future event) being higher than the normal value: exp1 > norm.

Joy, we can think of as the current situation being above average, regardless of past expectations for the current situation, or expectations for the future based on current stimuli. Therefore, exp1 and exp2 are not relevant here and there is only the discrepancy of act > norm.

Anger may seem a difficult emotion to classify in these terms, as it is generally regarded in relation to its associated response (attack) rather than to preceding events. However, it is still possible to imagine events that would typically lead to an angry response: being robbed, stubbing one's toe, being undeservedly humiliated. In all of these cases, a loss is involved that is perceived as unjustified and (should be) redeemable. In other words, the value of current events is perceived as below normal standards, but there is an expectation of improvement. In the case of undeserved humiliation, the value level of self-esteem could be currently reduced, but expected to be increased again when a suitable response is made. On a purely functional level, an active, aggressive response is pointless if there is no potential for improvement. In this case, the actual value would be smaller than normal. However, the first expected value exp1 (the expectation for the event in the future) would be larger than the actual value: act < norm and exp1 > act.

Sadness is, as quoted above, a "response to an event that has already taken place" (Barr-Zisowitz 2000). This emotion is strongly associated with irretrievable loss of some kind, where irretrievable could be taken as low expectations for the future. Thus, it combines a low actual value with an equally low or lower expectation for the future (exp1). This low exp1 value contrasts it with anger: act < norm and exp1 <= act.

Relief can be thought of as fear not being fulfilled: if the angry man holding the stick coming towards the fearful subject turns out to only be carrying some firewood, the subject will feel relieved. In other words, the expected value was below the normal value (as is the case in fear) before, but now it has increased towards the normal value again. As for the example above, where the stimulus of the approaching angry man with the stick evokes fear, and in the event (arrival of the man with the stick) nothing happens, the value that was expected for this situation before (exp2) is lower than normal, but the currently expected value (exp1) is not reduced (or less so): exp2 < norm and exp1 > exp2.

Disappointment, in contrast to relief, can be thought of as hope not being fulfilled, so a higher expectation than normal not coming true. This is analogous to relief as described above, which could be thought of as fear not coming true. In the example of the cooking smells emanating from the kitchen, disappointment would ensue if we then found that the food was being prepared for someone else: the value we were expecting initially (exp2) is above the normal level, but the value we are currently expecting (exp1) is lower than that, leading to the following conditions: exp2 > norm and exp1 < exp2.

Surprise occurs when we are not expecting anything, and a positive event occurs, despite our lack of expectation. This means the actual value is higher than normal, and higher than we were expecting previously: act > norm and exp2 < act.

To extend the range of emotions being considered, we introduce two further aspects. Firstly, there is known to be a separation in the brain between the regions coding for the future reward value of stimuli (as in the OFC) and for that coding for aspects of the self. These latter areas have been suggested to reside in the mid-brain region (Panksepp 1999). Thus, we consider a new set of value maps (likely sited in mid-OFC) for predicted values associated with self-esteem and rewards/penalties for the self. Secondly, there are predictions of rewards or

Table 14.2 Value comparisons and their associated emotions for different value types

	Reward/ punishment	Self-esteem	Outcome of action	Outcome of others action
exp1 < norm	Fear	Shyness	Guilt	Contempt
exp1 > norm	Hope	Confidence	Hope	Admiration
act < norm & exp1 > act	Anger	Anger	Regret	Empathy
act < norm & exp1 =< act	Sadness	Shame	Regret	Pity
exp2 < norm & exp1 > exp2	Relief	Gratitude	Relief	
exp2 < norm & act <= exp2			Guilt	Schadenfreude
exp2 > norm & exp1 < exp2	Disapp'tm't	Disapp'tm't	Frustration	Pity
act > norm & exp2 < act	Surprise	Surprise	Surprise	
act > norm	Joy	Pride, triumph	Satisfaction	Jealousy

punishments for future actions as guided by internal motor models in the parietal-premotor system (sometimes called mirror neurons but more correctly part of the overall motor system) (Raos et al. 2004, 2007). We suggest that there is also a separate set of value maps related to (predicted) action outcomes in the OFC. Such predictions can also be applied to another person, as seen by direct observation by the subject. The resulting set of value maps and associated thresholds leads to the enlarged set of emotions as displayed in Table 14.2.

As above for reward and punishment, now follows a detailed description of how each emotion in the above table is connected to the comparisons in the first column (when different from those in the second column described above).

14.3.1 *Self-Esteem*

Low or unstable self-esteem has been connected to anger and depression. As such, we hypothesise that positive changes in self-esteem have emotional consequences as well. In the literature, we can find a distinction between global self-esteem, self-evaluations and feelings of self-worth or state self-esteem. Global self-esteem is generally viewed as a personality trait determined in youth and adolescence (Brown et al. 2001), whereas self-evaluations refer to separate evaluations of the various abilities and characteristics of the self, and state self-esteem refers to temporary emotional states arising from positive or negative self-related stimuli. In this theory, we interpret global self-esteem as the normal value of self-esteem, and state self-esteem as the more variable actual and expected values of self-esteem, as these are responsive to incoming stimuli. As such, discrepancies between the innate global self-esteem value and the stimulus based state self-esteem values would give rise to emotions. Self-esteem is related not only to one's own judgment of the self, but also to that of others; it is assumed here to be a highly social concept representing our own assessment of our status, our ranking in the pecking order.

Particular emotions are especially related to the self, such as pride and shame. This has also been noted by other authors. They have termed this class of emotions 'self-conscious' (Lewis 2000) or 'self-referential' (Zinck 2008) emotions. Comparisons between different self-esteem value representations can lead to self-referential emotions in the following ways:

Pride could emerge from a stimulus giving rise to a high value of self-esteem, like a compliment or task well completed. This means the actual value of self-esteem is higher than the normal one: act > norm.

Shyness can be interpreted as a response to a threatening social situation in which we expect that our self-esteem could potentially be damaged. When giving a presentation, for example, a subject may feel shy and insecure if they have not prepared well and therefore expect to receive a dent in their self-esteem if it turns out not to be appreciated by the audience: exp1 < norm.

Confidence could be taken as the opposite of shyness, in the sense that it is a response to a social situation in which we expect our self-esteem to be increased. A subject would feel confident when they have prepared well for a presentation they are giving, and therefore expect the applause and compliments at the end: exp1 > norm.

Shame is strongly associated with public humiliation and a feeling of exposure. It could emerge in response to a reduced self-esteem when there is no expectation for this self-esteem to be redeemed: act < norm & exp1 =< act

Anger has been researched somewhat more thoroughly than other emotions as a consequence of changes in self-esteem values. Self-esteem has long been known to be involved in the development of feelings of anger. It has long been thought that it was low self-esteem per se that caused anger and aggression (Baumeister et al. 1996) but more recently it has been suggested that an instability of (Franck and De Raedt 2007) or/combined with threat to self-esteem (Kuppens and van Mechelen 2007) creates violent responses. Baumeister et al. (1996) also point out that it is not low self-esteem per se, but a discrepancy between a global high self-esteem and a more temporary perception of a disagreement with this self-assessment somewhere in the outside world that gives rise to anger. As explained above, this could be equated to a discrepancy between the normal and actual value of self-esteem, where a lower actual that normal value would give rise to anger. However, as mentioned in the 'Anger' section above, it seems likely that feelings of anger also encompass an expectation of a higher value in the future: act < norm & exp1 > act.

Gratitude may be a counterintuitive item on this list, as it appears to be by definition a response to receiving a gift, which is not directly related to self-esteem. However, gratitude does not necessarily need to be focused on one particular person and/or gift, as we can feel grateful in general for e.g. our good health. We propose that gratitude is the response to a stimulus increasing the actual value of self-esteem (e.g. a compliment) when the subject is feeling shy or insecure (so has a low previous expectation of self-esteem). Let us add that receiving a gift does not always lead to an experience of gratitude, and can be interpreted as a stimulus that increases self-esteem: exp2 < norm & act > exp2.

The reasoning behind the appearance of disappointment and surprise in this column in the table is similar to that applied previously for these emotions.

14.3.2 Outcome of Choice

Some emotions, like guilt or regret, are immediately connected to our actions or choices and their associated goals and motivations. If we define a goal or motivation as the expected result of an action or choice, we can consider this action to be a stimulus evoking a particular expected value (the goal). After all, when a goal-directed action is performed, this implies that the expected result of this action has some motivational value. The following emotions can all be related to choice motivations and expectations in this manner:

Satisfaction is the emotion that arises when we know that our actions have had a positive result, or in other words, that we made the right choice: $act > norm$.

Guilt can be related to a negative consequence of an action, but depends strongly on the intent of the individual when performing the action. As an example, let us take an action that most people would feel guilty about: throwing an egg at an innocent old lady. Regardless of whether or not the egg actually hits the old lady, it is the intent of doing something malicious that should already make us feel bad before the outcome of this action is known. What could reduce our guilty feelings is if the action was intended to have a positive effect (e.g. we were trying to hit the pickpocket who was trying to steal her bag at the time). In this case, our expectation was positive rather than negative. So we see that it is the expectation we have at the time of performing the action that makes us feel guilty (or not). In other words, if our expected value (as a consequence of an action) is lower than normal, we feel guilty: $exp1 < norm$. It should be noted that, contrary to fear, feelings of guilt continue if the outcome is indeed as negative as we expected: $exp2 < norm$ & $act < norm$.

Frustration is an emotion akin to anger. The difference between the two is that in frustration there is an element of thwarted goal achievement: our actions do not have the normal result, combined with an expectation that this state of affairs will be improved again in the future. For example, when a queue we are in is not moving as fast as we expected we get frustrated, while we still expect that (possibly aided by our actions of protest) we will reach the end of the queue at some point: $act < norm$ & $exp1 > act$.

Regret, the feeling associated with wishing that we had not made a particular choice in the past, is related to an action or choice that has had a negative result. Contrary to guilt, where the past action is the focus of the emotion and the outcome is irrelevant, in regret it is the outcome that matters, and the expectations at the time of the action are ignored. It is possible to regret an action without feeling

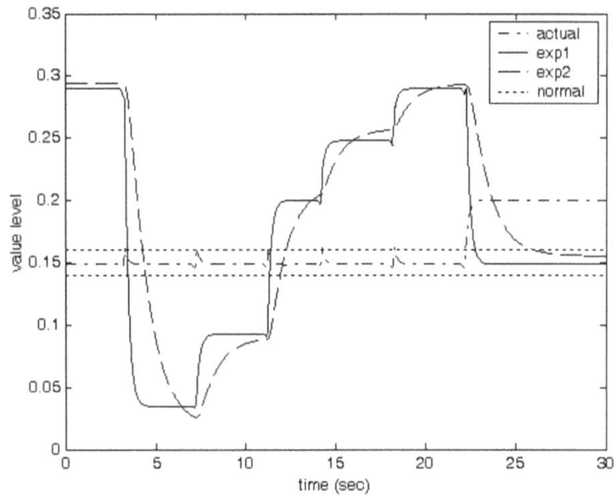

Fig. 14.3 Value levels for the 'sprinter' example

guilty about it, and vice versa. Let us suppose that in the earlier example where the egg is thrown at the old lady's bag thief, the egg accidentally hits the old lady instead of the bag thief. In this instance, the subject (the egg thrower) would probably regret their actions, but not feel guilty, as they were only trying to help the old lady. So regret is felt in cases where the actual value (the outcome of the action) is lower than normal, where our previous expectation (at the time of the action) is irrelevant. In contrast with frustration, there is no expectation of improvement of this outcome: act < norm & exp1 < = act.

To illustrate this range of value comparisons through action outcomes, let us consider an example of an emotion- evoking situation as in the example of the man and the stick above: a professional sprinter running an important race. This is a highly emotionally charged situation, where the emotional experience would depend at least for a large part on the expectation of the reward of winning the race, as a consequence of the sprinter's action of running it. The value changes involved in this example are represented in Fig. 14.3. Before starting, the sprinters would psych themselves up to be absolutely convinced that they can win the race (or, if there are opponents whom they are sure are much stronger, that they can achieve a different goal such as beating their more direct opponents). This would create a high expected reward value of the outcome of their action of running as fast as they can, where their normal value (average result in the past) may be to become fourth. Were they to get into the last position after a bad start, their expectations would be lower than normal, but as they overtake each of their opponents in turn, their expectation would rise back to the level of winning the race. Were they to be overtaken at the last minute and end in second place, their exp1 and actual values would both reflect this second position, whereas exp2 would remain slightly higher.

14.3.3 *Actions of Others*

Our evaluations of the actions of other people also play a role in the emergence of some emotions, like pity and jealousy.

Jealousy arises when another person's actions result in a disproportionate reward, or when they receive reward without having deserved it, in other words when their actions should not lead to a reward. The value of their possessions or wellbeing as a consequence of their action is higher than normal: act > norm.

Pity is created by the opposite situation: another person possesses less than they should, when their actions should have resulted in a higher reward or less punishment than they actually have. The value of their possessions or wellbeing, as a consequence of their actions, is lower than normal: act < norm.

Contempt is related to the likelihood that a person behaves in a way that is stupid, cowardly or otherwise not beneficial to themselves or others. When someone performs an action that we deem unwise or wrong, because we expect this action to result in a loss, we may feel contempt for this person: exp1 < norm.

Admiration, as the opposite of contempt, is related to the likelihood that a person behaves in a way that is clever, brave or otherwise beneficial to themselves or others. When someone performs an action that we deem wise or right, because we expect it to result in a reward or increase in wellbeing for themselves or others, we may admire this person: exp1 > norm.

Empathy is similar to pity, with the difference that in empathy we expect the subject to be able to cope on their own, and sympathise with their struggle, whereas pity suggests a victim that needs help: act < norm and exp1 > act.

Schadenfreude is an emotion that can follow contempt, if a person's unwise actions result in the expected punishment: exp2 < norm and act <= exp2.

14.3.4 *Other Emotions*

Some emotions have a prominent place in the emotion literature, but are missing from Table 14.2. Perhaps the most notable one is disgust, which has often been named as one of the basic emotions. It could be argued that disgust is a drive rather than an emotion, as it is strongly related to body state, similarly to other drives like pain and hunger. Disgust may constitute a direct physical response to food or body related stimuli that are best avoided. The more abstract meaning of disgust whereby one can indicate to be disgusted with for example a person behaving immorally could be taken as a linguistic analogy rather than indicating a particular emotion. In this case, the actual emotion may be contempt.

Other notable omissions from the above list of emotions are love and hate. Although it is impossible to deny that these are actually true emotions, they do differ from the emotions listed above. Love and hate are not so much produced by one particular episode, but tend to be longer lasting states, attached to one particular

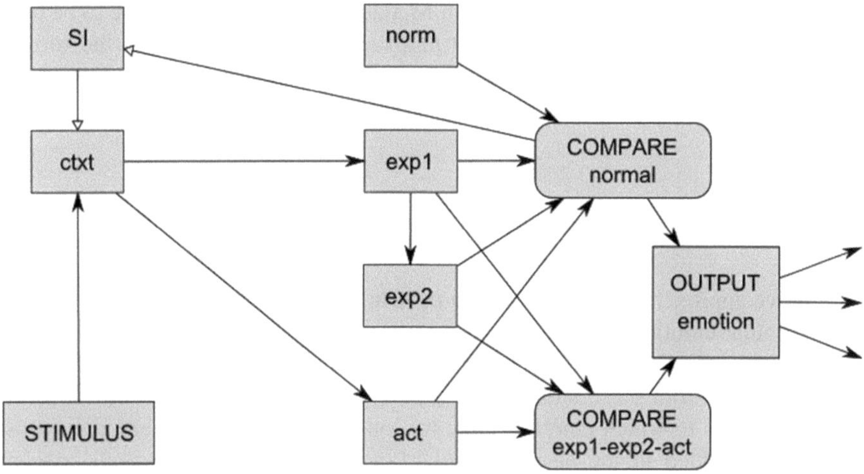

Fig. 14.4 Representation of the structure of the simulink model. In the 'COMPARE normal' module, act, exp1 and exp2 are compared to the normal value, whereas in the 'COMPARE exp1-exp2-act' they are compared to one another. Both signals are passed on to the 'OUTPUT' module, in which they are combined to represent the different emotions as per Table 14.2. Output from the 'COMPARE normal' module disinhibits the 'ctxt' module via the 'SI' module, maintaining a representation of stimuli that may no longer be present

individual; it is hardly possible to feel love or hate for a little while in response to a single stimulus. These emotions encompass not only a feeling, but also imply a commitment. Without this commitment, love is reduced to mere liking and hate is reduced to dislike, neither of which can be classed as fully fledged emotions, merely as evaluations.

A final emotion missing from the list above is embarrassment. Although it has been proposed that embarrassment is merely a milder version of shame, empirical study shows otherwise (Tangney et al. 1996): differences in ratings of subjects' embarrassing and shameful experiences cannot be explained by intensity of affect alone, although exactly how embarrassment differs from shame is still not clear.

A question that arises is how, in a computational framework like this, the intensity of the emotion could be determined. The cases of fear, hope and joy, where only one value comparison is involved, seem relatively straightforward: it would be the size of this discrepancy that determines the intensity of the emotion. For the other emotions, it seems likely that both relevant comparisons contribute in some way to their intensity. Let us, for example, look at sadness, which we feel when confronted with an irretrievable loss. The amount of sadness we feel would depend both on the current reduction in our actual value (the perceived loss) and our expectation for the future: if we expect things to get worse, we will be sadder than if we expect things to stay the way they currently are.

The comparisons as described above and summarised in Table 14.2 can be calculated through the neurally inspired computational architecture outlined in Fig. 14.4 (Korsten 2009).

This architecture was implemented in Matlab Simulink, and was able to be used to show how the temporal dynamics of many situations bringing forth the various emotions can be computed. We refer the interested reader to (Korsten 2009) for a more detailed description of these results.

14.4 Bringing the Emotions into Consciousness

The above approach to emotions can in principle apply to any situation, predicting the particular emotion that will ensue in response to that particular situation based on the value discrepancies implied by the context and stimuli perceived by the subject. Apart from these situation specific predictions, this model also contains implications as to which emotions may be considered basic; those emotions most generally included in the basic emotions (anger, fear, joy, sadness) all concern reward/punishment without involvement of the exp2 value. Furthermore, it is possible to predict which emotions can and cannot occur simultaneously for the same value type.

This theory has a strong interdisciplinary character; it can be expressed computationally, related to various psychological theories on emotion and is neurally inspired. Moreover, neural correlates for the model have tentatively been identified (Korsten 2009). Also, it accounts for a range of emotional processes, encompassing short prototypical emotional episodes as well as longer lasting states. Nevertheless, this model does not account for attentional or perceptual influences on emotion and vice versa, or for any subtleties in the emotional response based on response inhibition for instance. It is also unspecific as to how the emotional feeling – the perception of the emotion – arises.

As such, it needs to be considered as a proposal for a core process rather than a complete theory encompassing all aspects of emotional functioning. We now need to try to expand this core emotional framework so that emotions can be 'felt' – so become conscious. It is suggested here that the appraisal-based model of emotions we have presented will be able to function as a basis for the addition of attention and consciousness, the model of emotions being composed of modules (such as presented in Fig. 14.4) that are either outside attention or whose circuitry is not compromised by the addition of the attention control system as discussed in Chap. 5, with its extension to the CODAM model of consciousness in Chap. 6.

We must consider now how we can fuse, in one manner or another, these emotions so that a subject may or may not be conscious of them, but at least is in one or other of those two states. We know that in both states – conscious or unconscious – emotional states can occur. How could that be?

An important aspect of attention is its ability to be directed exogenously (from outside) so as to turn to a new target if it is salient enough. Thus we would expect the fusion between emotion and consciousness to be through a suitably strong emotional signal breaking through into the attention circuit to redirect attention to the emotionally salient new target.

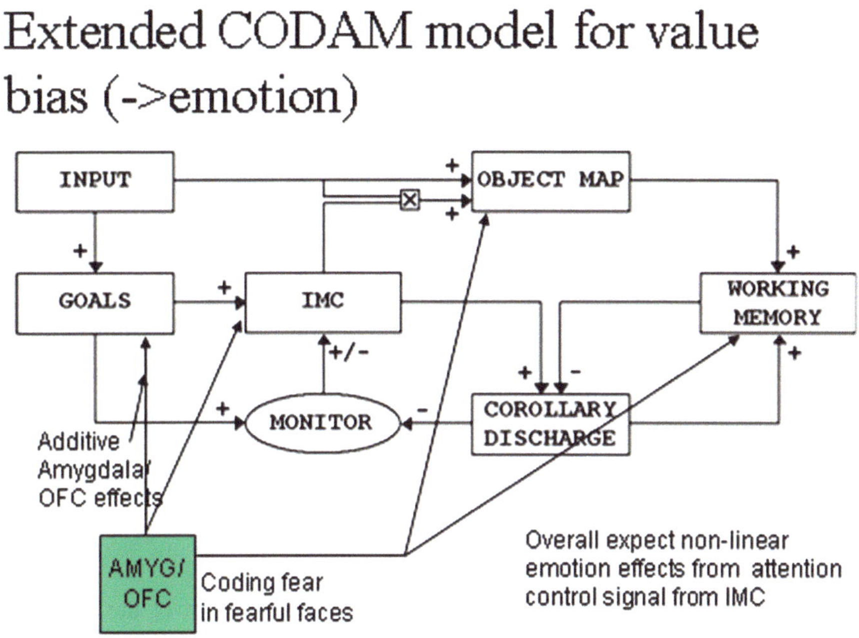

Fig. 14.5 A model of the mechanism by which emotions can interact with the CODAM-based attention control system. The AMYG/OFC module (composed of the amygdala and orbito-frontal cortex, with the predicted value maps used in the appraisal model of Sects. 14.2 and 14.3). The signals from AMYG/OFC are assumed to influence various of the modules in the CODAM architecture (specifically the Goals, IMC, object maps and working memory modules) so as to influence processing of various emotionally shaded stimuli

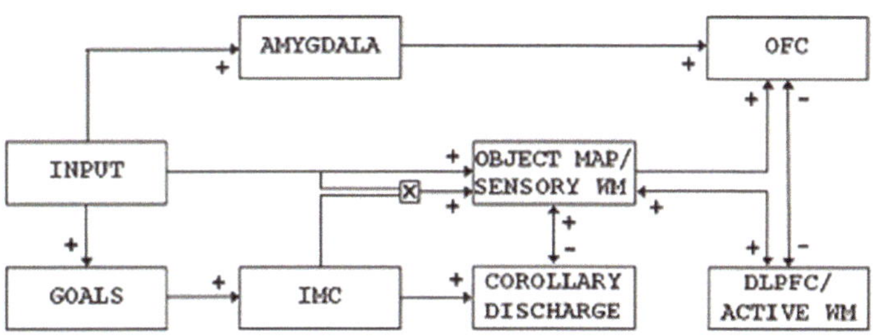

Fig. 14.6 Extended emotion model including PreFrontal cortex

Such a model has been constructed and applied to various fMRI-based paradigms (Fragopanagos and Taylor 2006), with the emotional signal being sent importantly to the attention control generator (IMC) in the CODAM model. Data was well fitted by the resulting fused emotion-CODAM model. The detailed architecture of the model is shown in Figs. 14.5 and 14.6 above.

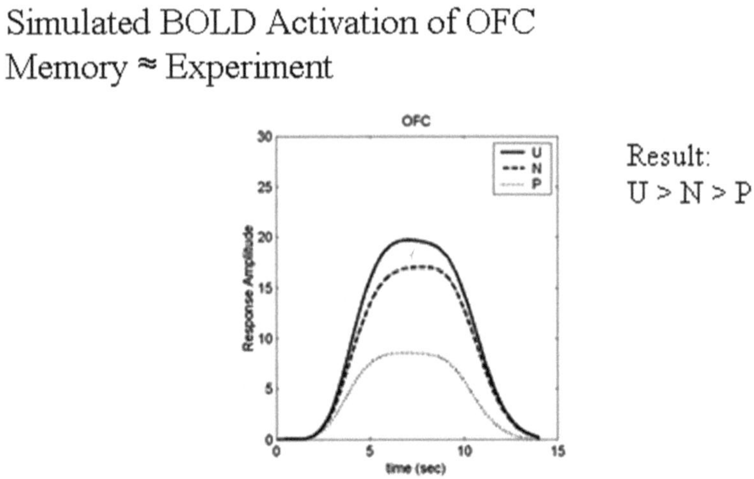

Fig. 14.7 The simulation of the memory task

This architecture was designed specifically to take account of the inhibitory interaction assumed present between the DLPFC and OFC/AMYG to explain the results of (Perlstein et al. 2002).

The paradigm of (Perlstein et al. 2002) consisted, as a detection task, of a probe stimulus consisting of a set of nine pictures, the task being to detect if the centre picture was identical to one of the eight pictures in the surround. These pictures, especially the central one, could be neutral, pleasant or unpleasant. A memory task was also run, with the stimulus to be remembered being either similar to or not of the target central picture shown later; again the pictures could be neutral, pleasant or unpleasant (N, P or U) (Fig. 14.7).

In the figure the three fMRI curves over time for DLPFC activation (for N, P and U stimuli) are in agreement with those obtained experientially by (Perlstein et al. 2002).

As noted in Fig. 14.8 there are numerous predictions from such an approach which have still to be tested.

However our task goes beyond such numerical data fitting; we want to be able to probe how it is that the actual conscious experience of an emotion can arise through whatever mechanism is appropriate. Presently our task is to fuse CODAM and the appraisal-based emotional framework we have just developed in this chapter.

Returning to the original appraisal architecture of Fig. 14.4, we consider such appraisal as arising from brain sites, such as in the orbito-frontal cortex, or more likely from a cingulate-related brain region. The cingulate area is thought to possess good connectivity to the parietal, so we assume that any emotional appraisal signal

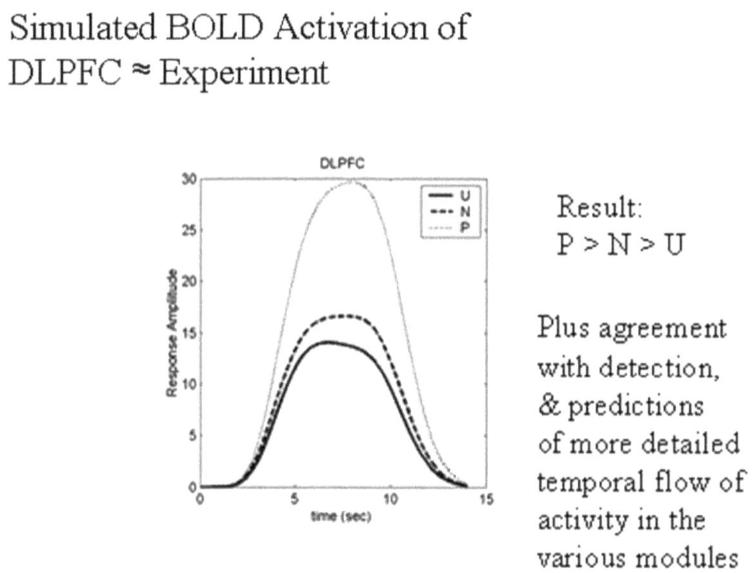

Fig. 14.8 Simulation of the detection task

is sent to the latter parietal region. Making the further assumption that some of this signal is sent to the attention movement signal generator (IMC) of CODAM then we additionally assume that this signal (arising from an emotionally salient stimulus), if large enough, will be expected to cause a change of the IMC signal to produce a redirection of attention to the new emotionally salient stimulus.

Such a redirection of attention will bring the content of the emotionally-salient stimulus into attention focus. At the same time we assume the corollary discharge of this new signal to activate further the relevant components of the appraisal circuitry of Fig. 14.4. These components will then be related to the stimulus content of the emotionally salient stimulus and thereby, we suggest, grant it an emotional flavor. We will 'feel' about the stimulus (love, hate, anger or whatever) as is appropriate, in a manner brought about by the concomitant activation of the appraisal circuit and of the sensory components of the stimulus. It is in this way that we can fuse emotion and consciousness.

Such an extension clearly requires considerably more experimental analysis than is presently available. However we hope that such investigations, but now with an emotional bias, can be applied to the quest for conscious emotions. Thus the use of emotionally-laden stimuli in the case of the experiments in (Robitaille and Jolicoeur 2006) could be followed up.

References

Anders S, Lotze M, Erb M, Grodd W, Birbaumer N (2004) Brain activity underlying emotional valence and arousal: a response-related fMRI study. Hum Brain Mapp 23:200–209
Armony JL, Servan-Schreiber D, Cohen JD, Ledoux JE (1995) An anatomically constrained neural network model of fear conditioning. Behav Neurosci 109:246–257
Barr-Zisowitz C (2000) "Sadness" – is there such a thing? In: Lewis M, Haviland-Jones J (eds) Handbook of emotions. Guilford Press, New York, pp 607–622
Baumeister RF, Smart L, Boden JM (1996) Relation of threatened egotism to violence and aggression: the dark side of high self-esteem. Psychol Rev 103:5–33
Brown JD, Dutton KA, Cook KE (2001) From the top down: self-esteem and self-evaluation. Cognit Emot 15:615–631
Fontaine JR, Scherer KR, Roesch EB, Ellsworth PC (2007) The world of emotions is not two-dimensional. Psychol Sci 18:1050–1057
Fragopanagos N, Taylor JG (2006) Modelling the interaction of attention and emotion. Neurocomputing 69(16–18):1977–1983
Franck E, De Raedt R (2007) Self-esteem reconsidered: unstable self-esteem outperforms level of self-esteem as vulnerability marker for depression. Behav Res Ther 45(7):1531–1541
Grimm S, Schmidt CF, Bermpohl F, Heinzel A, Dahlem Y, Wyss M, Hell D, Boesiger P, Boeker H, Northoff G (2006) Segregated neural representation of distinct emotion dimensions in the prefrontal cortex-an fMRI study. NeuroImage 30:325–340
Korsten N (2009) Neural architectures for an appraisal basis of emotion. PhD thesis, King's College London, London
Korsten N, Taylor JG (2010) A computational basis for the emotions. In: Gökçay D, Yildirim G Globa IGI (eds) Affective computing and interaction: psychological, cognitive and neuroscientific perspectives. Global IGI, pp 94–109
Kuppens P, van Mechelen I (2007) Interactional appraisal models for the anger appraisals of threatened self-esteem, other-blame and frustration. Cognit Emot 21:56–77
Ledoux JE (1992) Brain mechanisms of emotion and emotional learning. Curr Opin Neurobiol 2:191–197
Lewis M (2000) Self-conscious emotions: embarrassment, pride, shame, and guilt. In: Lewis M, Haviland-Jones J (eds) Handbook of emotions. Guilford Press, New York, pp 623–636
Ortony A, Clore GL, Collins A (1988) The cognitive structure of emotions. Cambridge University Press, Cambridge
Panksepp J (1982) Toward a general psychobiological theory of emotions. Behav Brain Sci 5:407–467
Panksepp J (1999) The preconscious substrates of consciousness: affective states and the evolutionary origins of the self. In: Gallagher S, Shear J (eds) Models of the self. Imprint Academic, Exeter, pp 113–130
Perlstein WM, Elbert T, Stenger VA (2002) Dissociation in human prefrontal cortex of affective influences on working memory-related activity. PNAS 99(3):1736–1741
Plutchik R (2001) The nature of emotions. Am Sci 89:344
Raos V, Evangeliou MN, Savaki HE (2004) Observation of action: grasping with the mind's hand. NeuroImage 23:193–201
Raos V, Evangeliou MN, Savaki HE (2007) Mental simulation of action in the service of action perception. J Neurosci 27:12675–12683
Robitaille N, Jolicoeur P (2006) Fundamental properties of the N2pc as an index of spatial attention: effects of masking. Can J Exp Psychol 60(2):101–111
Russell JA, Barrett LF (1999) Core affect, prototypical emotional episodes, and other things called emotion: dissecting the elephant. J Pers Soc Psychol 76:805–819
Sander D, Grafman J, Zalla T (2003) The human amygdala: an evolved system for relevance detection. Rev Neurosci 14:303–316

Sander D, Grandjean D, Scherer KR (2005) A systems approach to appraisal mechanisms in emotion. Neural Netw 18:317–352

Scherer KR, Schorr A, Johnstone T (2001) Appraisal processes in emotion: theory, methods research. Oxford University Press, New York

Schoenbaum G, Roesch M (2005) Orbitofrontal cortex, associative learning, and expectancies. Neuron 47:633–636

Suri RE, Schultz W (1999) A neural network model with dopamine-like reinforcement signal that learns a spatial delayed response task. Neuroscience 91:871–890

Tangney JP, Miller RS, Flicker L, Barlow DH (1996) Are shame, guilt, and embarrassment distinct emotions? J Pers Soc Psychol 70:1256–1269

Zinck A (2008) Self-referential emotions. Conscious Cogn 17:496–505

Chapter 15
Solving the Mind-Body Problem

15.1 The Mind-Body Problem

The mind is composed of mental fragments – sensations, feelings, thoughts, imaginations, all flowing now in an ordered sequence, now in a chaotic fashion. There are also non-conscious components involved in early brain processing of stimuli (as in lower level processing in vision, such as in V1) or in emotions not yet in consciousness (as discussed in Chap. 14). On the other hand the body is constructed to obey the underlying laws of physics, and its components obey the well-enumerated laws of physiology. It is these characteristic differences – between mind and body – that leads to the Mind-Body problem.

The Mind-Body problem has been in existence for several thousand years – going back to Plato, Aristotle, The Buddha and many other ancient Greek and Eastern thinkers. The problem is simple to state: the mind and the body seem to be entities of very different kinds (as just described), so how do they interact so as to produce in a person a mind able to have effects on their body (as when the person wills the body to perform some act), whilst the body can certainly strongly affect the mind (as in the case of an accident causing strong pain).

Although the problem is simple it has as yet no satisfactory solution, in spite of the amount of time and thought devoted to it over the ages.

Much has been written on the variety of solutions to the Mind-Body problem. There are the dualist solutions (mind and body are distinct, although then the problem of how they interact becomes even more embarrassing, and is so far unresolved in any satisfactory manner); the idealist solution (there is only mind, with body/matter being merely a manifestation of mind). How mind could have created the beautiful subtlety of the unification of the nuclear, electromagnetic and radioactive forces of nature, with a proposed extension (as yet unproven) to fusion of those forces with the force of gravity through superstrings, seems impossible to comprehend. A third position is that there is only body, so this is a purely reductive physical approach (but then the problem of how our amazing mental world, full of our 'mental' experiences are thereby created from matter, has to be faced). There

J.G. Taylor, *Solving the Mind-Body Problem by the CODAM Neural Model of Consciousness?*, Springer Series in Cognitive and Neural Systems 9,
DOI 10.1007/978-94-007-7645-6_15, © Springer Science+Business Media Dordrecht 2013

are also nuanced versions of one or other of these three initial positions for solving the Mind-Body problem, some having great sophistication and subtlety. But none has yet gained universal acceptance.

There is also the question of what exactly is the mind? It is certainly composed of conscious components, but it would also seem to contain non-conscious ones as noted earlier, such as unconscious emotions and low-level processing as yet out of consciousness; automatic motor responses are also below the radar of consciousness. These unconscious or pre-conscious components are not problematic since they do not possess apparently non-material *private* components like those that conscious components appear to do. A stone is not considered as having such private experiences as do we; it's responses to action on it can be described in purely physical terms. Similarly the non-conscious processes in our minds can be accepted as arising solely from suitable brain processing, so more easily understood as components of the body (especially with the great advances in brain science tracking down and modeling in detail this pre-conscious dynamic neural activity). These non-conscious components of mind do not have (by definition) any conscious component, so there is no difficulty in expecting them to arise solely from brain activity: they can therefore be seen as on the body side of the mind-body duo.

Thus the truly inexplicable part of the mind – that part to which the epithet *Mind-Body problem* most closely applies – appears to be that of consciousness. That position is appropriate to take here because it addresses what appears to be the most difficult part of the overall problem of the mind-body interaction: how conscious experience can interpenetrate and fuse with bodily activity so that they affect each other in the way mentioned in the first paragraph. Detailed neural models of the other (non-conscious) components of brain activity are increasingly convincing such as for predictive value coded by dopamine, for motor responses guided by internal motor models, and for early models of visual and other sensory processing through a hierarchy of increasingly complex feature detectors; these brain-based activities clearly support the lack of any *non-conscious mind*-body gap. These features support the approach used here, and are in line with modern thinking on the Mind-Body problem, especially that brought to the fore by Descartes, and since emphasized in the notions of the *hard problem* (Chalmers 1996) and the *explanatory gap* (Levine 1983): it is the gap between consciousness and brain activity that now constitutes the Mind-Body Problem. This chapter proposes to give a simple explanation of that gap and how it can be closed, based on earlier developments in the book.

15.2 The New Knowledge Base

The knowledge base we can bring to attempt to solve the mind-body problem has changed enormously over the thousands of years of the problem's existence. Matter has been probed down to very short distances through the development of ever higher energy particle accelerators, so that we now have an understanding of matter in terms of the gauge theories unifying the electromagnetic and weak forces and further fusion with the strong nuclear forces by means of the quark-gluon gauge theory. Together with gravity, understood through Einstein's general relativity, the comprehension of matter has improved down to distances of more than 17 orders of magnitude. In 2011 this distance decreased even more by the coming on-stream of the Large Hadron Collider, the LHC, at CERN. Hopefully it will find support for the basic features of the present level of unification, in terms of the discovery of the Higgs boson (as the basis for mass in the Universe), as well as providing us with new clues as to the next steps of probing ever deeper into the material world, such as by finding supersymmetric partners (or even supergravitational partners) to the present spectrum of *elementary* particles (for which there are even now claimed *bumps* in particle processes best explained, it has been claimed by their discoverers, by these exotic possibilities).

This progressive and impressive unlocking of the secrets of matter has only recently begun to be caught up by new ideas and new tools in brain science. These have allowed an ever better understanding of the manner in which the brain creates and controls experience. This knowledge is allowing ever deeper appreciation of how defects in brain processing brought about by various sorts of damage to the brain can enormously modify the overall experience of the person. There is still a large distance to travel to bring the knowledge of the brain up to the level of our enormously more detailed understanding of matter. But that gap appears to be closing fast, with a possible hiatus in the physics of the elementary particles due to the presently intractable character of the mathematics of superstrings (the only way known to sensibly unify gravity with the other forces of matter); no such hiatus appears in brain science except, of course, for the mind-body problem. Will that be the rock on which neuroscience founders in a similar manner, it is claimed by some physicists, as that of unifying the forces of nature will similarly founder on the rock of full unification of the forces of nature by inclusion of gravity with the other three forces?

The tools of brain science have been put to good use in probing the brain both locally and globally. The machines of PET and fMRI, measuring changes in blood flow in local regions of the brain due to neural activity, and EEG and MEG, measuring the electric and magnetic activities related more directly to neural activity in the brain, are allowing an increasingly precise view of the networks active across a range of functions. Combined with trans-cranial magnetic stimulation, the timing of the flow of crucial activity across the functional networks in the brain is beginning to give up the brain's secrets. At the same time correlated activities of small groups of single neurons in different regions of the brain are indicating how these various regions process activity at a local level, but also as part of a larger network of such regions in interaction. Ideas from dynamical systems theory have begun to help us understand better the brain at this intermediate mesoscopic level. In all, brains are becoming ever more understandable across all scales.

The major problem arising from this for the various idealist solutions outlined in the previous section to the mind-body problem are now clear: how can such precise physicalist explanations as have now been developed of the physical world at both micro and global scale be obtained from a purely mental universe? Mind does not enter into any of the detailed physical models (such as Einstein's Theory of General Relativity, or the Standard Model of elementary particles based on quantum field theory, with the increasingly well-explored and experimentally justified $SU(3) \times SU(2) \times U(1)$ symmetry). The problem is as to how to derive such well-established theories from a mental approach to the universe. As far as this author knows, there is not even a single framework available to attempt to solve this problem, nor a single idealist paper discussing it.

15.3 What is Is?

There is a further very important problem about the mind side of the Mind-Body problem which the enormous advances of high-energy physics and brain science (described above) have exacerbated, weighing the scales ever more heavily on the side of matter: How can any scientific discussion ever be given of an entity which is non-material? This question is not one to be taken lightly. If the response is that in any case science is not the relevant arbiter here, then the whole discussion will be unable to make any forward progress to resolve any of the important issues:

1. How can mind and matter interact in any dualistic theory. The difficulty here is that there are no comparable features of mind to those of matter. Thus where are quantitative descriptors such as energy, particles, waves, or related quantities present in detail in descriptions of matter, down to quantum fields describing the creation and annihilation of multiplets of the most elementary of particles. How to match these up to corresponding descriptors of immaterial entities conjectured as composing mind?
2. How could such a theory be tested in any quantitative manner if there is no such quantitative theory?
3. Only by appeal to the world of experience can we conclude an argument between opposing positions. If it is claimed that part of that experience is in any case purely mental, so above the scientific arena, we can only point to the powerful developments in brain science that depend on inner report of subjects but ties them down to the activities of specific parts of the brain (or their absence in the case of deficit studies). Thus inner experience is used in a scientific manner here. Even the distortions of schizophrenics and those with other mental diseases are presently being used to probe those brain components, such as the attention system, which are damaged in mental diseases (as in schizophrenia, discussed in Chap. 10), and may help explain the distorted experience reported by sufferers.

At this point it is relevant to attempt to define a non-material entity. The entity must be such that it has no material features, or at least not all its features are required to be material. But that leads us to the need to define a non-material feature.

We can attack the problem of defining a non-material feature in a negative manner. Thus a non-material feature is one that cannot be considered as existing in the material universe: it can have no space or time dimension or possess any energy or weight. More completely it cannot be acted on by any of the forces of nature. This definition has clarity, but is it possible to turn it into a positive definition in terms of the presence of some basic ‘immaterial attributes’? On further reflection this appears very difficult, for any non-material attribute has no possible description in present language. Thus it would appear we cannot even begin to define a non-material entity.

At this point it is usual for the interlocutor to step back and say that science has not observed all possible aspects of the Universe. ‘There are more things under heaven and earth, etc, etc’. But there are not – that claim is completely untrue – if these entities are of a material nature. The properties of matter – its decomposition into ever smaller components down to quarks, gluons and W & Z particles – all fit together so remarkably accurately down to those incredibly short distances of at least 17 orders of magnitude (and now more, in the LHC) that there can be nothing material that has been missed. Granted dark matter and dark energy are still not understood, although they could fit well into the new *zoo* of supersymmetric particles I mentioned earlier that may be discovered at the LHC, and in any case were only involved at a very early stage of the Universe. They are not relevant to us here and now at the sorts of energies and sizes of objects we can normally deal with.

The simplest aspect of the problem of even discussing a non-material entity or feature is that it could possess no possible attributes enabling it to be compared to anything in the material world. It has no shape or length, no weight, no color, or any other attribute that can be compared with any physical entity. In other words it can only be made of indescribable attributes. It could be claimed that it must therefore be somehow composed of purely abstract attributes. But what could these be? Even purely mathematical symbols must have some quantitative ground to be made sense of. Moreover one would be falling into a Platonic trap by assuming that a mathematically defined entity actually exists in the world. In reality it only exists in the brains of mathematicians who have worked on it. But that existence is in terms of the geometrical or algebraic objects codes used by those mathematicians to think about their concepts.

Abstract concepts such as *love* (used much in attributions to God in religions across the world) are themselves suspect as to their meaning. If such a concept is to be used in relation to our human interpretation then it is quantifiable (*I love you very much*), due to its basis in bodily processes and behavioral responses to the loved one. It is thus not immaterially based. Recent developments in appraisal approaches to emotions and how these appraisals might be created by suitable brain activity (Korsten 2009; Korsten and Taylor 2010, along with many other researchers, some referenced there), indicate strong support for a brain-basis for all human emotions (as developed in Chap. 14).

If an immaterial entity has no relation to any material entity then the names ‘entity’ or ‘immaterial ‘do not help. It can be claimed by the proponent of the existence of immaterial entities that such entities exist but we can never have any knowledge of them – as in the claim that ‘God is ineffable’. But then we end up with attributes which can only be defined outside our experienced universe, so completely outside our ken. If that is so, why should they even be said to exist. For the claim of ineffability, of existence always outside our ken is one which gives us no information at all. The immaterial entity is forever unknown. It may as well not exist. Any claim that it does exist but we will never know it, is just putting off the day of reckoning. The immaterial entity does not exist, as far as we can ever know.

Nor can so-called supernatural entities fill the bill. My own studies in the arena of the supernatural (Taylor 1980) indicated there are no such entities which can be detected by modern experimental techniques. All the experiments reported in (Taylor 1980) indicated that there were either no effects (except thosc brought about by conscious or unconscious cheating) or that there were those that could be explained by relatively simple normal physical laws (electrostatics, neurophysiology, etc.). Thus the supernatural avenue is a dead end as far as immaterial entities are concerned.

We arrive at the conclusion that it is difficult to analyze non-material entity in any manner. Given that the entity cannot be described in any quantitative manner at all, nothing can be said about it that can be agreed or disagreed about with any possibility of truth (where by the latter I mean scientifically checking up on the statement). Non-material entities or components are thus very slippery ones which twist and turn between your fingers so strongly that they escape you before you have had a time to investigate them in any reportable manner. But if you cannot report on the nature of these entities then you cannot even usefully discuss them. Even with yourself. But if you cannot do that then such entities cannot be part of any useful conversation (with others or with yourself): no progress can ever be made in probing them.

It might be said that there must be something outside this material universe, since it itself began only some 13 or so billion years ago in a big bang. What was present before that? It can't have been anything material, since the material universe did not exist before then. Could the universe have been created somehow from an immaterial entity?

This immateriality could have existed in the multi-dimensional universe that may have been present before 13 billion years ago (as in a superstring universe in 10 or 11 dimensions). But then it would not actually have been properly immaterial. We know too little of the physics of the whole universe presently, however, to be able to extrapolate effectively before the first 10^{-40} of a second of the Big Bang. So we cannot presently discuss that aspect in any scientific detail. But in any case those very early times (pre-10^{-40} of a second) would still be in the material universe, so the claim of ignorance is not going to help us bring in immaterial entities in interaction with the material universe. No model exists to help us understand how there could be a phase transition, say, from immateriality to materiality just at the time of the Big Bang. So we cannot bring in immateriality by the back door, so to speak, into the material universe.

It might be claimed that transcendental entities, like the wave function of the Universe or the transcendental ego of Kant are themselves immaterial. But the former of these is the basis of all measurements on the Universe or its component parts, so has many observable derivations. It may be said to be the basis of the observable quantities of the Universe, the Ur-ground of the possibility of observing anything in the world; it also cannot be used to bring in immaterial entities. Nor can the second of the above items – the transcendental ego of Kant (or the pre-reflective self of Husserl and colleagues) – since that again has directly observable aspects in terms of human behavior and report. It is the 'I' which has been the main topic of this book so far.

The overall conclusion of this is that idealism and dualism face difficult problems of definition of the basic components – immaterial entities – which presently have not been solved. We thus turn to physicalism, which has its own hard problems, and most particularly the *hard problem* of providing a detailed explanation of how mind can be created by suitably subtle dynamics of matter (Chalmers 1996) and of crossing that seemingly unbridgeable gap between mind and matter (Levine 1983). This is still unsolved in a universally accepted manner, but various promising approaches have been developed over the last few decades. These involve the reduction of the mind to the dynamics of brain activity involving some, if not all, of the billions of nerve cells in the brain. We will consider this at a functional level, so do not explore building up to that global character of the brain from the lower single nerve cell level (which, however, is increasingly effectively being attempted). So we turn to look at possible global principles for the brain – its crucial functional components.

15.4 Further Analysis of the Mind

- In the previous section we arrived at the disturbing conclusion that 'immaterial mind' is not a concept which can be grounded in any known scientific manner, so is outside the universe of scientific discourse. Since several billion people on the earth would seem to have solved the problem of talking about mind, through a variety of religious faiths involving an immaterial mind, we cannot treat the result of the previous section without some further discussion.
- It was shown that 'mind', regarded as the immaterial component of our own experience, would never seem to be a concept available for scientific investigation. That is a conclusion to be expected, since we define science as a method to analyze the material world around us, and definitely not the non-material world, whatever that is. But what is this supposed further component of the universe? Let us turn to consider how it could be approached.
- Our main thesis in this chapter is that we can solve the Mind-Body Problem in a simple manner, through the very simple material reductionist method: all is matter. Such a claim would seem to be immediately countered by that just stated: the mind will never be available for scientific investigation. To resolve this dilemma we need to expand the notion of science to include some aspects of personal experience. In any case that has already been done in many neuropsychological experiments, as already noted, where inner report is crucial to understand the results. In fact in many such experiments inner report is part

of the 'scientific' results. So the notion of science has thereby been expanded by default to include mental activity and its results.

- As an example of such an expansion, consider an experiment claimed to be testing the relation between attention and consciousness (Rahnev et al. 2009). In it a subject tries to detect two sets of stimuli: those at the centre of the field of view and those in a peripheral position. In order for the central viewing to be paid most attention (due to the difficulty of some centrally-placed task), the peripheral task then becomes a very difficult one; the peripheral task might be such as detecting the presence or absence of a dimly illuminated shape. Then the simultaneous detection of a peripheral figure will be a difficult additional burden on attention. Yet results seem to indicate that such a burden can be handled by consciousness of the peripheral figure. So apparently consciousness and attention have thereby been separated by such an experimental paradigm (but see Chap. 3 for a rebuttal of this claim). Crucially the experiences of the subject are being taken into account as part of the results of the experiment.
- Given that such an expansion of the notions of science as purely objective to include inner private experience has occurred, then further account of inner experience must be allowed. Where this stops is unclear, although experiments taking account of numerous subjects, not just depending on the claims of a single individual, would seem to have the best chance of being veridical; the accounts of single individuals, such as those with blind-sight, are good examples as to where controversies can arise due to the actual experiences of single subjects and how they are best to be interpreted in an object view.
- Even if we accept much in the nature of inner report, we still have to go back to the brain and its dynamical activity in order to understand, from the physicalist position, how such inner report arises in the first place. So we need to consider if we can discover general principles of how the brain might function to produce such inner report.

15.5 Global Principles of the Brain's Action

An important part of understanding how reportable activity could arise purely from brain activity is that of the extraction of basic principles of brain dynamics at a global level to help comprehend the awesome power possessed by these coupled circuits of local nerve cells. Several global principles can be extracted from the welter of possibilities. Three of these are arguably the most important:

- Attention, as a control system to filter lower-level brain activity so as to allow very few stimulus representations to enter the higher-level arena of thought and manipulation of neural activities; these filters are mainly controlled by activity in parietal and pre-frontal cortices. Attention thereby allows the higher level processes such as thinking and reasoning to work on a much smaller number of stimulus representations. Without such filters we would be swamped by the 'blooming buzzing confusion of the external world', as William James put it. These aspects are discussed more fully in Chap. 5 on attention and its extension to consciousness in Chap. 6).

- Emotion, in terms of value maps learnt in orbito-frontal cortex (and also coded in associated amygdala sites) so as to bias what is to be processed and to guide choice of task goals (by their associated predicted rewards). These value maps, fused with body activations and automatic brain-stem responses, are used to give emotional value to decisions for action. More has been said about emotion in Chap. 14, especially in terms of the appraisal approach and its possible computational basis in the brain.
- Long-term memory created on-line, so as to allow for incremental wisdom being acquired about the environment for use as a guide for further interactions.

Other global principles can be added to these three possibilities, such as the use of hierarchical processing (as noted especially in vision, so as to create flexible visual codes for complex objects which can be used at a variety of scales). Another possibility involves the use of synchronization of neurons over long distances, such as at 40 Hz and other frequencies, so as to solve the binding problem of combining the different codes for objects (as occur in multi-modal hierarchical coding schemes described earlier). A further principle is that of recurrent loops of neural activity, to allow for the creation of short term memory (working memory) sites for the temporary holding of such activity for spreading around to other similar sites, so acting as 'report' centers in the brain.

15.6 Global Principles of the Mind's Action

In order to make progress on how the conscious mind is created by brain activity, we must also outline what it is we consider is the main structure of the mind. This is a highly controversial area, with many possible suggestions. However a certain degree of clarity seems to have arisen from recent ideas about the bipartite nature of consciousness by Western phenomenologists (Zahavi 1999, 2005b, 2006). From this increasingly influential body of writing, consciousness can be seen to be divisible into (1) Content. This is available for report and internal manipulation, being composed of the neural activities (codes) representing the colors, hardness and so on of objects, for example of the specific scent of the rose and the taste of the wine, and more generally of the various modality-based features of external objects in our consciousness; (2) Ownership (for further discussion on this aspect of consciousness see Chaps. 7 and 8). This second component is termed by Western phenomenologists as the pre-reflective or inner self. The inner self is to be regarded as that denoted by the singular *I*. The inner self is supposed to have no content, the term *pre-reflective* implying that it is at a more primitive level than any reflective descriptor of oneself, such as possessing a beard or moustache or being blond. The pre-reflective self is required to be present in the mind as an independent entity along with content in order to prevent an infinite regress which would otherwise arise when considering *I* as defined by reflection on one's characteristics (Shoemaker 1968). Thus although the pre-reflective self is content free it is an essential companion of the content of experience: without an owner there can be no content, no-one to have the experience. This aspect has been especially emphasised by the French philosopher Merleau-Ponty (Merleau-Ponty 1945) where he noted that the word *consciousness* has no meaning independently of the pre-reflective self. Sartre (Sartre 1943) also pointed out that the pre-reflective self constitutes the very mode of being of experience. It provides the bedrock of 'what it is like to be' in terms of ownership of the contents of experience; *I* am the experiencer (Nagel 1974). Such a view is also supported by Western philosophers such as (Flanagan 1992) who points out that a form of low-level consciousness (identifiable with the pre-reflective self) involves crucially experiencing an experience as 'mine' (so as being the owner of the experience). A further aspect of the pre-reflective self accepted by these thinkers is that it is a constant and immediate component of consciousness, present "whenever I am living through or undergoing an experience" (Zahavi 2005b, p. 1; ibid 2006).

The manner in which these two components of consciousness – content and pre-reflective self – can interact is highly problematic. It was not resolved by the phenomenologists, who seemed to take the majority view that the pre-reflective self arose from the body. That is a controversial position, with the danger that subjects who have lost proprioception should lose their inner self; this is known not to happen, so making that position suspect (Cole and Paillard 1995). Also sufferers from Cotard's syndrome (where the subject claims they are dead or missing their blood or internal organs, in association with depression or suicidal tendencies) provide further contrary empirical evidence (Metzinger 2003). Such patients have at least loss of a body image and of emotional content (related to their depression) but more extremely may no longer use *I* to talk about themselves: they seem to be losing their pre-reflective self. This appears to be independent of the loss of body image. This question is also discussed more generally, in terms of experimental evidence for or against the existence of a possible preflective mental state involving proprioceptive feedback, in (Taylor 2007). However in (Damasio 2000) evidence is presented that loss of the body image does cause a loss of phenomenal consciousness. Yet any body-based self crucially does not have the automatic property of 'immunity to error through misidentification of the first person pronoun' (Shoemaker 1968), regarded as a crucial component of consciousness. The phrase denotes the fact that I cannot sensibly ask you 'Are you sure it is you in pain?' when you tell me you are feeling pain: you just know it is you. Thus the problem of understanding how content and owner can interact is still open, and in some ways can be seen as hard as that of explaining consciousness more generally. However it has reduced the task to something more specific in terms of the supposed two components of consciousness.

Yet the problem of understanding how the owner of conscious content becomes *aware* of that content is decidedly non-trivial. This is partly because the owner or *I* is itself not properly understood. It is possible to consider the nature of *I* as expounded in the writings of Eastern meditators, who have spent decades meditating on the nature of their inner self and have built up a vast literature on their experiences across a range of disciplines, such as Zen. However this material, whilst potentially very valuable, is also solely descriptive and highly personal, thereby making it difficult to relate to modern brain science. The increasing number of brain imaging experiments performed on those who are meditating is gradually helping to bridge this gap, so that this vast literature from meditators could ultimately help exploration into the inner self. This process was also helped by the work of (Varela 1996) who gave an initial attack on the brain basis of meditation; a further discussion is given in (Taylor 2002a, b), where an attention-based neural model was proposed of the pre-reflective self and how it could explain the mechanism of the pure conscious experience reported by meditators. This subject is also considered in Chap. 13.

This bipartite division of consciousness is now increasingly being accepted as an important avenue to attack mental disease, especially schizophrenia (Sass and Parnas 2003; Taylor 2010). The distorted mental experiences reported by patients and occurring as a major part of the disease are now being seen, through a broad study of many sufferers, as arising from an imbalance between these two components of consciousness (content and the inner self). Thereby an important brain basis for this two-component approach is being built, through increasingly detailed analysis of the accounts of schizophrenics (as developed in Chap. 10), which should considerably strengthen and test the bipartite understanding of consciousness.

Another influential bipartite division of consciousness was introduced in (Block 1995). This proposed the two components of consciousness as being phenomenal consciousness and access consciousness (P and A for short respectively, as used by Block). P consciousness is just experience, granting the subject the experience of 'what it is like to be' (Nagel 1974). Thus a P-conscious state is an experiential state. One function of P consciousness was suggested as integrating the outputs of specialized sensors and feeding the results to the executive system. On the other hand, A-consciousness involves states that are poised for use as a premise in reasoning and for rational control of response.

There has been much separate debate of these two different bipartite approaches to consciousness, although little attempt known to the author to compare and contrast them. It is clear that they are quite different ways of slicing up experience. Thus both P and A consciousness possess content, so that these have been fused together by the Western phenomenologists. However the *owner* of experience posited by the Western phenomenologists does not appear in its present 'no-content' guise anywhere in the P-A split of (Block 1995). The latter would thus seem to be vulnerable to the infinite regress of (Shoemaker 1968). It is also to be emphasized that the owner or pre-reflective self may best be regarded as agency-free (Taylor 2002a, b, 2009). Thus the owner cannot initiate any action, except that of standing as the guardian at the gate to consciousness, to prevent any incorrect intrusion of a distracter into consciousness and so provides the important feature of 'Immunity to error through misidentification of the first person pronoun' (Shoemaker 1968). This phrase describes the feature raised before, that I know if I am in pain; you cannot ask me "Are you sure it is you who is in pain?"

The question of the relation of ownership to agency needs to be considered further, since the experience of agency ('did I cause the movement or did he/she?') and its possible defects have been seriously proposed as the basis of schizophrenia, for example (Frith 1992). This appears to contradict the suggestion made earlier (Chap. 10) that such deficits arise from an imbalance in the relation between the owner and what is owned (the content). There have been several useful discussions of agency in action, such as in (Gallagher 2000), and its possible breakdowns as might appear in schizophrenia. If however there is an attention basis for the owner/content consciousness system, then there is no agency here. This is known from discussions going back to William James. As Bricklin pointed out (Bricklin 1999, p92) "Nothing in James' paradigm suggested that the predominance of consciousness of one thought over another was generated by the power of attention; rather the predominance in consciousness was itself attention".

This lack of agency of attention can be seen more clearly by the experimentally-based deconstruction of the attention control system into a goal-biasing system guiding the movement of the focus of attention, itself generated by a further control module, sending a signal back down into lower level posterior cortices. This bias is what causes the movement of attention and that itself is fed by values of goals, stored elsewhere. Moreover, as pointed out in (Taylor 2002a, b) there is the added danger that if there were agentive experience in attention movement then there would be danger of an infinite regress. This would arise on exploring the source of the consciousness attributed to the attention mover.

In conclusion there is an important difference between agency ('I caused it') and ownership ('it was my thought'). Both may break down in schizophrenia, leading to a range of possible symptoms and distortions of experience (Taylor 2010 and Chap. 10). These are thought to be explicable along the Western phenomenological division of consciousness (Sass and Parnas 2003); a similar explanation along the P-A division is not presently available.

15.7 Ownership and Its Importance

A range of brain-based solutions have been put forward to solve the Mind-Body problem, which were discussed in Chap. 4, other than CODAM. This itself was developed in Chaps. 5 and 6, and then the ownership aspect developed over Chaps. 7, 8 and 9.

There are numerous important features from the models of consciousness described in Chaps. 4, 5, 6, 7, 8 and 9 that we must take forward to build a more complete model with some chance of success: attention as a filter and possible source of feature binding, activity coming into awareness when accessing sites of working memory, associated attention copy *ownership*, various forms of memory as crucial components to give content to awareness, synchronization as a possible additional source of feature binding for object representations, suitably complex interacting brain states, emotion values as giving the whole system value and source of drive, and so on.

Ownership has been emphasized in various publications (Taylor 2002a, b, 2007, 2011a) and discussed in Chap. 7. What is ownership and how can we probe it further? It has been already pointed out in Chap. 7 and various places noted above by the author that ownership is a subtle aspect because it involves neural activity which is coded at a high level, in particular to remove distracters from the focus of attention, and has no or very little connection with the lower level content codes of sensory cortices in the brain. It is this divorce from content which leads to the 'nothingness' claim of Sartre (Sartre 1943) and the difficulty to detect the ownership experience itself. But if there is an experience of ownership (as claimed by the phenomenological school of Husserl et al. (Zahavi 2005b)) then what is being owned and who is doing the owning? These are questions deserving a crisp answer before we can embark on further search for that owner.

The answer to the question as to what is being owned is that it is the content of the experience that is about to arise into reportability across short-term memory sites, as in the GW model of (Baars 1997). This content is itself being worked on, as in the N2pc and the SPCN discussed in the previous section and especially in Chap. 9. Various distracters are being removed from lower level processing or from the visual short term memory (VSTM) (in the case of vision) before report can be achieved. It is this cleaning–up of the results of earlier processing, carried out before content awareness arises, which is what ownership appears, from this data, to be about. The cleansing process itself grants to the subject the experience of preparing the content to allowed to be reported, thus casting the stamp of ownership on it.

Besides the purely 'academic' cleaning process there may also be goal bias signals arising from emotional/reward sites (say in orbito-frontal cortex) or from competitive discriminative support from cingulate regions. At the same time memory processes or prefrontal regions may be activated to bring about rehearsal of earlier material, as needed by rehearsal to guide further actions. Thus a number of other regions could contribute to the ownership feeling. However we suggest that the primary ownership content may well arise from the basic activity of distracter removal. Indeed that is the purpose of attention itself: to allow for filtering out distracters and concentration on the brain representation of target activity in order to be able to handle further processing of only the latter. That is all that attention consists of: target search and related inhibitory processes to remove distracters.

Do these various components of brain activity begin to look as if they could provide a basis for a suitably scientific theory of consciousness? This can only be achieved by a modeling-cum-experimental attack on the many paradigms and associated observations in which consciousness crucially enters, and based on models including the three main principles mentioned earlier and some or all of the features of the various models outlined above. Such phenomena include, for example, the attentional blink, the Stroop test, attention search and associated phenomena such as the Posner benefit paradigm, processing of subliminal inputs, the phenomenon of blindsight and numerous others. Only when all of these have been explained satisfactorily at a quantitative level by such a developed model can we begin to be able to justify the claimed final solution to the mind-body problem. Even then there will be enormous problems of interpretation.

We have already (in Chap. 9) given some experimental support for the CODAM model. We have also shown in Chap. 7 how the CODAM model is in agreement with the results claimed in Western phenomenology. We have shown how various meditative states (in Chap. 11), as well as the distorted states arising in schizophrenics (Chap. 10), can be understood by CODAM. Thus CODAM seems to have stood up to all of the present tests. There are indeed many, many further tests. But CODAM has passed all its present ones.

15.8 The Solution to the Mind Body Problem

Our solution to the Mind Body Problem is thus as follows:

1. We approach the problem from a purely physicalist approach. In other words we have no immaterial entities; their existence is very dubious, as we discussed earlier in this chapter. However such a solution is not new, but thousands of years old. What do we bring to the table that is new?
2. We already have considerable support for the CODAM model of consciousness of Chap. 6. We discussed it from various angles in Chaps. 7, 8, 9, 10 and 11, with positive results throughout (as noted at the end of the previous section). In other words we now have both an experimental and theoretical basis for the CODAM model of consciousness, it being based purely on the activities of the brain (subtle as they may be).

It is the second component that we claim is new, in supporting the physical solution to the Mind Body Problem we have presented in this book. Without an explanation as to how consciousness can arise from the activity of a purely physical brain the physicalist solution to the Mind Body Problem is not necessarily any better than the other two solutions (the idealist and dualist approaches). Each involves levels of belief. Each has no scientific proof.

However once we have such a viable solution to the problem of the creation of consciousness from purely physical activity in the brain, then the situation is very different. Now the problems (especially the lack of any way to approach immaterial entities) of the idealist and dualist solutions, mentioned earlier in this chapter have to be accepted as truly serious if not completely life-threatening. On the other hand the physicalist solution now has both real value as well as opening up an enormous area of human activity to a completely different view from heretofore. The soul is now part of the body, although a very subtle one at that. Humans are but machines, although again very subtle ones. So we have to look again at humans both as isolated machines as well as in societies of such machines. The human world is thus very different from that which many conceive it to be. It is much simpler (it does not contain ineffable entities, about which nothing can be said) but it is much, much subtler. It is undoubtedly a thing of great beauty and complexity when we are included in it in a correct and satisfactory manner.

References

Baars B (1997) Some essential differences between consciousness and attention, perception, and working memory. Conscious Cogn 6(2–3):363–371

Block NJ (1995) On a confusion about the function of consciousness. Behav Brain Sci 18:227–247

Bricklin J (1999) A variety of religious experience: William James and the non-reality of free will. In: Libet B, Freeman A, Sutherland K (eds) The volitional brain. Imprint Academic, Thorverton

Chalmers D (1996) The conscious mind: in search of a fundamental theory. Oxford University Press, Oxford

Cole J, Paillard J (1995) Living without touch and peripheral information about body position and movement: studies with deafferented patients. In: Bermudez JL, Marcel AJ, Eilan N (eds) The body and the self. MIT Press, Cambridge, MA, pp 245–266

Damasio AR (2000) The feeling of what happens: body, emotion and the making of consciousness. Vintage Press, New York

Flanagan O (1992) Consciousness reconsidered. MIT Press, Cambridge, MA

Frith CD (1992) The cognitive neuropsychology of schizophrenia. Erlbaum, Englewood Cliffs

Gallagher S (2000) Philosophical conceptions of the self: implications for cognitive science. Trends Cogn Sci 4(1):14–21

Korsten N (2009) Neural architectures for an appraisal basis of emotion. PhD thesis, King's College London, London

Korsten N, Taylor JG (2010) A computational basis for the emotions. In: Gökçay D, Yildirim G, Global IGI (eds) Affective computing and interaction: psychological, cognitive and neuroscientific perspectives. Global IGI

Levine J (1983) Materialism and qualia: the explanatory gap. Pac Philos Q 64:354–361

Merleau-Ponty M (1945) Phenomenologie de la perception. Editions Gallimard, Paris. English translation: Smith C (1962) Phenomenology of perception. Routledge and Kegan Paul, London

Metzinger T (2003) Being no-one. MIT Press, Cambridge, MA

Nagel T (1974) What is it like to be a bat? Philos Rev 83:434–450

Rahnev D et al (2009) Inattention boosts subjective visibility: implications for inattentional and change blindness. J Vis 9:157

Sartre J-P (1943) L'Etre et le neant. Gallimard, Paris. English translation: Barnes HE (1956) Being and nothingness. Philosophical Library, New York

Sass L, Parnas J (2003) Schizophrenia, consciousness and the self. Schizophr Bull 29(3):427–444

Shoemaker S (1968) Self-reference and self-awareness. J Philos 65:555–567

Taylor JG (1980) Science and the supernatural: an investigation of paranormal phenomena including psychic healing, clairvoyance, telepathy, and precognition by a distinguished physicist and mathematician. Copyright 1980 by P. N. Taylor. E. P. Dutton, New York. State "First Edition" with number line: "10 9 8 7 6 5 4 3 2 1"

Taylor JG (2002a) Paying attention to consciousness. TICS 6(5):206–210

Taylor JG (2002b) From matter to mind. J Conscious Stud 6:3–22

Taylor JG (2007) On the neurodynamics of the creation of consciousness. Cognit Neurodyn 1:97–118

Taylor JG (2009) Beyond consciousness. Int J Mach Conscious 1(1):11–21

Taylor JG (2010) A neural model of the loss of self in schizophrenia. Schizophr Bull 37(6):1229–1247
Taylor JG (2011a) The problem of I. J Conscious Stud (in press)
Taylor JG (2011b) The corollary discharge of attention exists? Conscious Cognit (in press)
Varela FJ (1996) Neurophenomenology: a methodological remedy to the hard problem. J Conscious Stud 3:330–359
Zahavi D (1999) Self-awareness and alterity. Northwestern University Press, Evanston
Zahavi D (2005a) Being someone. Psyche 11(5):1–20
Zahavi D (2005b) Phenomenological approaches to self-consciousness. Stanford Encyclopedia of Philosophy (online)
Zahavi D (2006) Subjectivity and selfhood: investigating the first person perspective. MIT Press, Cambridge, MA

Chapter 16
Is There Free Will in CODAM?

16.1 The Nature of Free Will

Is each of us free to act in whatever way we wish at any time? Or instead are we completely controlled by our brain processes, so that there is no such freedom? Given the CODAM framework for supposedly explaining all of our conscious experience, how is our strong sense of having such freedom an illusion? How could such a strong belief arise? If not, where is the gap in a CODAM to let the freedom of will occur? These are questions to which it is natural to expect answers. The sense of freedom of the will is strong in all of us. We are often told "It is up to you" about some decision we have yet to make. We naturally expect to have this sense of freedom explained by any detailed model of experience.

Many have tried to reconcile this feeling of freedom with any physicalist model of the mind. More generally there has been considerable discussion by philosophers of the nature of possible free will or reasons for its absence, over the last two millennia. We refer to others to describe the present situation in its various aspects (Strawson 1998, 2011; Dennett 2003). We assume here that determinism is valid, as physicalism dictates. Hence our task is to explain, by CODAM in particular, how the *illusion* of free will could arise in the deterministic brain.

There are numerous claims, such as by Dennett (2003) with respect to the chaotic and probabilistic nature of the world, which would seem to undermine determinism. However the analysis of the brain we have given so far does not involve any details of the manner in which the brain's dynamics is carried out. Undoubtedly there will be forms of chaos in the underlying time variations of at least some of the sub-sets of neurons, as well as the need for a quantum description of processes at the molecular level and below. But the level at which we have discussed attention in Chap. 5 and CODAM in Chap. 6 is purely at the classical level, where determinism (modulo chaos) is king. Thus we can expect that the above claim that the CODAM brain is deterministic is correct for our discussion of free will.

J.G. Taylor, *Solving the Mind-Body Problem by the CODAM Neural Model of Consciousness?*, Springer Series in Cognitive and Neural Systems 9,
DOI 10.1007/978-94-007-7645-6_16,

16.2 Making Up One's Mind in CODAM

The CODAM model we described in Chap. 6 has, as a crucial part of its architecture, an error-assessing module, termed the monitor module in Fig. 6.4. It is this module that is used to correct an error if the attention signal movement generator is predicted, by use of the corollary discharge, to be in error. It is in this way that the corollary discharge is used for early error correction. This module, for example, was employed in the simulation of the attentional blink (Fragopanagos et al. 2005). We identified the brain siting of this module as in the cingulate cortex, following many studies (Magno et al. 2006; Carter et al. 1998; Botvinick et al. 2004; Scheffers et al. 1996; Coles et al. 2001; Turken and Swick 1999; Brown and Braver 2005).

It is to be expected that the cingulate will be involved in decision-making. Indeed many studies have shown activity in this region when decision making is involved in an experimental paradigm. This is to be expected, since there are strong connections from the prefrontal cortex to the cingulate; it is in the prefrontal region that goals have been observed to be stored, as well as activating regions there by exogenous inputs of strong salience (Dias et al. 2006). Moreover there are also strong inputs to the cingulate from the reward regions in the orbito-frontal cortex, suitable value-based bias to putative goal inputs from the prefrontal cortex.

We assume, from this connectivity and the decision making activation of cingulate, that the cingulate plays an important (if not final) role in decision making. Such decision making activity is not clearly linked to lower level feature analysers, as occurs in the development of content consciousness. Thus, as in the case of the non-contentful nature of the experience of the inner self, there is expected to be a similar lack of any conscious content associated with the high-level decision making that is ongoing in cingulate and the associated activity in prefrontal and orbitofrontal cortices.

Here we meet two possibilities as to the nature of the experience associated with decision making. One is that it does actually become conscious, with the pros and cons of the various possibilities being consciously adumbrated and weighed up. Such conscious decision making is not of interest to us, since it cannot be directly involved in the experience of free will, for in such 'free' cases, decisions arise as if from nowhere. It is this 'nowhere' aspect that grants the feeling of freedom in the decision-making process. It occurs as if by magic. A machine could do the same if the pros and cons were being assessed consciously, with weights attached to the various possible decisions, and that decision being taken with highest weight. Thus we can assume that there is no conscious experience of the decision making process.

The processes mentioned above involved in decision-making in the cingulated clearly contribute to the experience of free will. Of course it would have been possible for someone to attempt to make purely random decisions in their head, so as to avoid any bias from possible rewards or other sources (from salience, for example). But it would be difficult for such a person to ensure there was no influence from the cingulate, prefrontal cortex or orbitofrontal cortex. Nor, of course, would one normally say that such random decision making would qualify under the aegis of free will.

We can summarise this section as the conclusion that:

Based on CODAM, free will is an illusion

Behind the illusion is the unconscious decision-making machinery as we considered above. This machinery provides the bias need to make up the mind, this bias being outside any conscious experience.

16.3 Free Will and Morality

How can the above explanation of the illusion of free will be consistent with any form of moral behaviour? And how can society provide any way of developing moral subjects? These are crucial questions for society, but ones for which we do not seem to have any clear answer. If free will is an illusion and a person's responses in a given situation are completely determinated by certain of his brain processes, then how can anyone be morally responsible for their actions? This is a serious problem for the law, but even more so all actions in a given society. How can we deal with such a serious and difficult world view as provided by CODAM, with free will as an illusion and no-one being responsible for their actions? Only brains are responsible, and persons are not.

This is clearly the nub of the problem: where is the person in their brain? CODAM has deconstructed the person so completely that it has disappeared, and only its fragments are left. There is no legal entity able to be regarded as morally responsible for an action. This is a problem that is being discussed increasingly in the literature (Dennett 1991; Klemm 2011).

The most important feature left out of this discussion is that of the plasticity of the brain. In childhood the brain is like blotting paper, able to soak up and memorise new experiences so that they become part of the memory of the subject, and so can guide it in future actions. In particular, from an early age the parents of an infant should be inculcating it with teaching the things that it can do and those it cannot. Penalties associated with the word NO (and rewards related to YES) must be embedded in the infant and child brain so it can function effectively in society and the emerging adult is able to be able to move equably around amongst its fellow humans.

It is at this stage where the basics of morality need to be taught. This process of learning needs to be continued into the adult, although the rewards and penalties will then have to be imposed by society itself in terms of the law and of the possible rewards from success in whatever area of endeavour the subject attempts.

If a subject, considered adult, then commits a criminal act, this is an indication that not all of its earlier upbringing had provided it with the right set of penalties and rewards. Re-education at the right level is then required, either in a prison (if the subject is too dangerous) or in a suitable environment if not. It would be in that way that society can achieve a composition in which the moral code of the society is preserved and the members of society are able to live together in some degree of equanimity.

16.4 Conclusions

There are many conclusions that may be derived from the above framework of the loss of true free will but the preservation of a moral society (of whatever form); many of these are discussed in my earlier book (Taylor 2006). We have already talked about the most important: how to inculcate moral behaviour in society's members. There are many other more detailed aspects raised in (Taylor 2006), such as the treatment of the mentally ill, the manner in which education should occur, how infants should be reared, the nature of the law and moral responsibility, etc. But these will have to be left to the interested reader. As expected, if the key to the nature of consciousness is uncovered, then it is expected that the whole structure of society will be seen in a new light.

References

Botvinick MM, Cohen JD, Carter CS (2004) Conflict monitoring and anterior cingulate cortex: an update. Trends Cogn Sci 8:539–546

Brown JW, Braver TS (2005) Learned predictions of error likelihood in the anterior cingulate cortex. Science 307:1118–1121

Carter CS, Braver TS, Barch DM, Botvinick MM, Noll D, Cohen JD (1998) Anterior cingulate cortex, error detection, and the online monitoring of performance. Science 280:747–749

Coles MG, Scheffers MK, Holroyd CB (2001) Why is there an ERN/Ne on correct trials? Response representations, stimulus-related components, and the theory of error-processing. Biol Psychol 56:173–189

Dennett D (1991) In: Lane A (ed) Consciousness explained. The Penguin Press, London

Dennett DC (2003) Freedom evolves. Viking Press, New York

Dias EC, McGinnis T, Smiley JF, Foxe JJ, Schroeder CE, Javitt DC (2006) Changing plans: neural correlates of executive control in monkey and human frontal cortex. Exp Brain Res 174(2):279–291

Fragopanagos N, Kockelkoren S, Taylor JG (2005) A neurodynamic model of the attentional blink. Brain Res Cogn Brain Res 24(3):568–586

Klemm WR (2011) Neural representation of the sense of self. Adv Cognit Psychol 7:16–30

Magno E, Foxe JJ, Molholm S, Robertson IH, Garavan H (2006) The anterior cingulate and error avoidance. J Neurosci 26(18):4769–4773

Scheffers MK, Coles MG, Bernstein P, Gehring WJ, Donchin E (1996) Event-related brain potentials and error-related processing: an analysis of incorrect responses to go and no-go stimuli. Psychophysiology 33:42–53

Strawson G (1998, 2011) Free will. In: Craig E (ed) Encyclopedia of philosophy. Routledge, London

Taylor JG (2006) The mind: a user's manual. Wiley, Chichester

Turken AU, Swick D (1999) Response selection in the human anterior cingulate cortex. Nat Neurosci 2:920–924

Chapter 17
Beyond CODAM?

17.1 The Story of Consciousness

We started this book with comments on the story of the Universe (Big Bang and all) and continued with only a brief part of the story of consciousness and its urgent need for completion at, for example, approximately up to the same level as the story of the material Universe. Are we now in a position to begin to complete the story of consciousness to that level, as we had hoped, from what has been said in this book? In particular, are the fragments of the story put together in Chap. 12 on the evolution of consciousness enough to allow us to complete the story? Have we reached the end of the road as far as consciousness is concerned?

There have been ideas, such as the existence of a Singularity in the development of machine intelligence to the point that it surpasses human intelligence, when it would be expected that humanity's consciousness will even begin to whither away; this has been strongly promulgated by Ray Kurzweil and colleagues, for example. Such a singularity would indeed be expected to cause some sort of transition in human society, although exactly what is highly debatable.

Another aspect relevant to this question of the story of human consciousness is that of possible alien consciousnesses? If they exist then they should have their own stories of development. Can we make any guess, guided by CODAM, as to the possible nature of these stories?

We have to take careful note (if we want to develop ideas of alien consciousness along CODAM lines) that CODAM has been crafted for the animal (and more specifically human) brain, based on the living nerve cell and its connections and inner structure. Should we not thus restrict our considerations to such specific alien control systems which are similar to the human brain? However that will clearly be limiting.

An alternative is to see what functionality is crucial in the mechanisms of CODAM, to be able to create consciousness. This would hopefully allow us to generalize beyond possible alien brains made of similar units to our own brain cells to those of very different structure, but possessing a similar functionality to

J.G. Taylor, *Solving the Mind-Body Problem by the CODAM Neural Model of Consciousness?*, Springer Series in Cognitive and Neural Systems 9,
DOI 10.1007/978-94-007-7645-6_17, © Springer Science+Business Media Dordrecht 2013

CODAM. It is that second route we will follow here, both because it allows us to look more deeply into CODAM as well as make more effective conjectures about the whole range of conscious architectures.

At the same time there has been a rapid progress in the attempt to develop machines with consciousness. This has centred mainly on creating programs that are self-monitoring, so can correct their own errors and achieve considerable success in a variety of intelligent tasks. However the resultant 'self' being considered here would seem (at least in most cases) to be of the reflective form, corresponding to looking into the mirror and finding (or unexpectedly not finding) the expected feature reflected back to you. It does not seem to correspond to any sort of 'I' along the lines we have developed in the last few chapters from Chap. 6 onwards.

17.2 Consciousness Guided by CODAM?

What, then, are the crucial functional components of CODAM that have made it be effective in creating our consciousness? We claimed in Chap. 5 that the whole basis of human, and we supposed animal, consciousness was that of attention (having given a more complete discussion of that position in Chap. 3). If we want to develop more fully how CODAM works across such a broad range of animals, it might be most useful to specify attention in a functional manner, rather than considering the specific neural details of the DAN-VAN network and others from primates, as we did in Chap. 5.

We have already, in various places in this book, specified the crucial and basic properties of attention: that of amplification of neural tissue coding for an attended target and inhibition of neural tissue coding for distracters. How these various amplifications and inhibitions arise is not, it would seem, crucial to the working of attention, provided it does work. Where the biases arise from (as in the case of the prefrontal goal structures we posited) is also unimportant. What is crucial is that the codes for targets are amplified and those for distracters inhibited by attention in a biased manner.

This is not a trivial process to achieve, of course. The codes for the various objects in an environment will in general have to be learnt; it would be impossible to provide detailed genetic code for the detailed set of objects in a standard complex environment. Even on somewhat degraded surfaces like that of the moon there is still a variety of objects (of all sorts of shapes and sizes and colours) to be observed and possibly exploited. So attention to this variety of targets in almost any complex environment will require considerable learning, as part of the learning process of the total range of objects.

A being may decide or be unable to make such attention selection of the objects in its environments. It will then be severely disadvantaged over other beings who do have such selective powers as provided by attention. That disadvantage arises since the being will not be able to concentrate on any particular object of interest, with removal of the (possibly many) other objects in its field of view (or of other sensory organs), for further cognitive-level processing. In other words it has a 'blooming buzzing confusion' at all levels in its sensory field. And it will therefore be highly disadvantaged when competing for food, etc against an attention-equipped being able, for example, to single out its enemy from the rest of its sensory field, or its easiest prey.

This then, is the advantage for a being with an attention apparatus of the CODAM form.

But the next step, to introduce consciousness, is one which has given us humans on earth an enormous boost over other animals. Can we describe that in a functional manner as we did for attention?

17.3 The Functional Form of Attention-Based Consciousness

We have to consider carefully the dynamics of the attention processes in the CODAM model. The basic feedback signal for re-directing attention is sent in CODAM from an attention movement signal generator module (term the inverse model controller or IMC in Chap. 6) to sweep back down the hierarchy of feature-processing modules so as to amplify the neural activities representing the to-be-attended stimulus while at the same time, it is supposed, there is part of this backward signal which is inhibiting the neural representation of any distracters in the input (where possible). That seems to be the functional basis for primate attention processing (see more details in Chaps. 5 and 6).

A general alien being, it is assumed, will have a basic feature-based input analysis as well as the feedback attention control signal of similar functionality to that of primates. This assumption is supported by the evolutionary competitive processes undergone in the evolution of the overall sensory systems in primates (and lower animals). It is assumed that similar competition for resources would have been met in the evolution of the sensory system of an alien being. There may be other solutions to that for the optimal input analyzing system, as we see here on earth in birds and in invertebrates. But without some form of hierarchical feature analysis system it is clear that the combinatorial explosion of trying to code the majority of the objects in a complex environment will prevent a satisfactory handling of the details of that complex environment.

Thus the attention feedback system will need to be able to handle amplification or inhibition of an aggregate of modules forming part of the hierarchy of the sensory system of the alien being. It is here, especially arising from the expected limited speed at which this feedback can travel, that we need to bring in a corollary discharge of the attention feedback signal. Such a corollary discharge can speed up the access of the codes for the attended target (however these codes are represented) by support of such codes in the ultimate buffer that they will access, as well as inhibit the codes of distracters which may cause confusion in any further processing.

Thus we come to a CODAM-style of processing, in which there is early processing at a high level associated with amplifications, and inhibitions are at work in the buffer to achieve the code for the attended target on its own. Such a buffer is to be expected also to exist, since without it there could not be effective processing at further levels (of Global Workspace form).

Speed-up, increased accuracy and similar features are thus at the basis of the employment of the corollary discharge component of CODAM. It is these features that help any being to live more efficiently in its environment. But then the argument developed in the earlier chapters, from Chap. 6 onwards, indicates how this corollary discharge would lead to an experience of 'nothingness' (due to the high level coding of the corollary discharge signal and its recipient sites). We have consciousness, with both its content and its 'inner self'! Such may be the case for many types of alien being.

17.4 To Be or Not to Be Conscious?

In the previous section we developed a functional form of CODAM which indicated the possibility in some alien beings of a form of consciousness somewhat similar to our own. We must now consider in what way an alien being would not employ or possess such an attention structure to allow us to expect it to be conscious like us. Indeed we should also consider what other forms of consciousness there might exist in such alien beings.

To begin this exploration it seems reasonable that we still expect the being to possess the basics of attention control. If not then its environment will indeed be a 'blooming buzzing confusion', so making it very vulnerable to other more sophisticated beings it might meet. In particular it would be very unlikely it could make contact with others, such as ourselves.

Given it has a basic attention power, we should consider this initially to be driven exogenously, so from outside. Such an animal as the crocodile would seem to have such attentive powers, driven to lunge towards any possible prey it detects in its environment. But it is difficult to see how such processing can lead to any consciousness of the response or the environment. Certainly there is no consciousness for the crocodile, no inner self. It is not needed and indeed would get in the way of the crocodile being a mean lean eating machine.

For endogenous or internal goals in control of attention, we would expect suitable buffers in the alien's processing system to enable it to bias the direction of its attention exactly to a desired attention target. Thus we expect an alien being's information processing circuitry to be similar, in general functionally, to that of the extended ballistic control circuit of Fig. 5.1 in Chap. 5.

However we still have to ask: do all alien beings require an additional corollary discharge component of the attention control signals? Would that grant them greater efficiency in their attention control or not?

In all the cases we have considered up to now in this book, even including the alien being in a complex environment, such an extra corollary discharge component will undoubtedly put them on the top of the pile. However if the environment of the alien being is rather simple – with rather separated targets and distracters, for example – then there may not be much gained, if anything, by having the extra corollary discharge to clean up distracters, to remove them rapidly so as to prevent slowing of information processing. Yet a being existing in a simple domain like that will not necessarily be prepared to handle more complex domains when it steps out (or flies out) into other environments, most of which would be expected to be as complex as our own here on earth.

We also need to consider the case of aliens living in domains far more complex than our own. Then most likely the most successful beings would have evolved to possess highly complex sensory apparati, which would themselves require considerable further processing to analyse and make use of the complex signals entering their sensory processing areas. But then we expect there to be a greater need for the alien to possess a corollary discharge apparatus as part of its attention system, so as to speed up and make more error-free the results of its attention filtering process.

17.5 Super-Consciousness?

The title of this last chapter is 'Beyond CODAM?' In other words could there exist a form of consciousness far more powerful than our own – a so-called 'super-consciousness'? If it could exist, what form might it take?

It is, of course, difficult to conjecture about remote and possibly very difficult aspects of an alien being's much more advanced processing than our own. We ended the last section with a suggestion about the powers of an alien in a more complex environment than our own. But what if that environment was even far more complex than our own? Would the alien have to evolve some sort of super-consciousness to be able to properly handle such a situation?

This is an important question to consider: what would such a super-consciousness be needed for? It is to be expected that a form of consciousness similar to that of CODAM would have already evolved, so the alien being would already have had considerable control over its environment. Would there have been a need to solve ever more complex equations describing that world? But then we would expect the alien beings to have developed computers to be able to solve computationally hard mathematical-type of problems about their environment. Presently it is unclear what a superconsciousness could do other than solve increasingly deep philosophical-type problems about the world. But then these are going to be similar to those we have been considering here, in this book. So why need superconsciousness?

Let us go back to more mundane but realistic possibilities. We can see that a very much more complex environment would require a correspondingly far more powerful sensory analysis system than we possess. For example there may be sensors across a far wider electromagnetic spectrum than one we inhabit. These enhanced sensors would produce many more codes at a high level for object classification, as well as an expected increase in the motor codes and associated affordances that would allow more efficient motor control over such objects. However that seems to require greater capacity for the sensory, classification and motor systems, but possibly not anything new in principle beyond what we possess ourselves. The same can be said for possible internal models to be employed in planning and reasoning by the alien so as to be as efficient as possible in achieving goals. In other words it appears there is nothing new here in terms of the standard processing of the alien compared to ourselves.

Is there anything new that might arise in the way the attention system of the alien being evolves in its supercomplex environment? Could a super-consciousness arise here? Of what would it consist?

Firstly we might consider if a hierarchy or an ensemble, of corollary discharges might arise? These possibilities could occur, for example, if the attention feedback control system itself was split into a set of feedback signals. However it is assumed that the activities in the separate lower level feature modules still arise from a single attended object, so that all of these separate signals would have to be co-ordinated so as to correspond to amplifying the various features of the attended object information representation. Such co-ordination would thus allow the generation of an overall corollary discharge signal, required to inhibit distracters. The distracters themselves would also be represented in a set of modules. However the crucial action of the corollary discharge would be expected to be at the higher short-term buffer level, and so correspond to that same level in CODAM. Hence a similar sort of inner self would arise as in that model. Nothing new!

Is it possible that in a highly complex environment there would be a necessary splitting of the corollary discharge signal so that it would be able to handle a plethora of such distracters? For example if there were four separate corollary discharges, each sent to a separate area on the short-term buffer coding for the various areas concerned? Would there then be four separate inner selves?

I think not necessarily. For each of those four corollary discharges needs to also carry information about the attended target representation, since otherwise they could inhibit that as well as distracters. Moreover each of the separate corollary discharges may also have to carry the same information: inhibit all distracters (without expectation that the distracters in a given area belong to a specific list). Thus the corollary discharges are to be considered as identical and there would only be one inner self.

We can only conclude, in this brief foray into the possibility of there existing superconsciousness in some alien beings, that in fact it seems difficult to see how such superconsciousnesses could have evolved, as well as being unable to arrive at a situation in which it is essential. Of course one could conceive a super-being effectively with a number of heads, each with its own consciousness. There is still a problem for the efficiency of such a being since it would need to have some sort of overall leader – and so back to a CODAM style of overall controller.

So far we have not investigated Pure Consciousness in this scenario. Is it possible that such a pure form of consciousness as the Pure Conscious Experience discussed in Chap. 11, or its further variants, such as Oceanic Consciousness (Forman 1990) could be of relevance here? In oceanic consciousness, there is supposed to be a melding of one's inner self with the sensory input from outside (Forman 1990). The nature of such a state (if it exists) is not clear, so it is difficult to conjecture on how this state could play the role of a superconsciousness. If it could then it has already, so it is claimed, been achieved by humans.

17.6 The Future of Consciousness

The somewhat negative results we have arrived at in the previous section do not imply that human consciousness (or other animal consciousness, if it exists) could not evolve further in the future. But such evolution would not obviously be associated with a new functionality, in other words with a 'super'-CODAM. The possible evolution of our consciousness, from the results and discussions in this chapter, would be expected to arise from the greater complexity of the object code regions of our brain, together with possible more complex motor actions on this wider range (and more complex) objects. In other words such evolution would produce an increase of complexity of the content of our consciousness, but not necessarily of our 'inner selves'. The 'I' itself could be enlarged by an increase in ancillary inputs (from emotion, long-term memory, etc, as discussed in Chap. 7). At the same time there would be an associated increase in the complexity of the corollary discharge signal due to the increased capacity of the lower level sensory and motor control systems. But the inner self would still be a 'nothingness', and possibly even more difficult to allow for meditation into a state of pure consciousness (due to the greater capacity of the various sensory and motor components acting as distracters).

Reference

Forman RKC (ed) (1990) The problem of pure consciousness: mysticism and philosophy. Oxford University Press, New York

Index

J.G. Taylor, *Solving the Mind-Body Problem by the CODAM Neural Model of Consciousness?*, Springer Series in Cognitive and Neural Systems 9,
DOI 10.1007/978-94-007-7645-6, © Springer Science+Business Media Dordrecht 2013

MIX
Papier aus verantwortungsvollen Quellen
Paper from responsible sources
FSC® C105338

If you have any concerns about our products,
you can contact us on
ProductSafety@springernature.com

In case Publisher is established outside the EU,
the EU authorized representative is:
Springer Nature Customer Service Center GmbH
Europaplatz 3, 69115 Heidelberg, Germany

Printed by Libri Plureos GmbH
in Hamburg, Germany